Among Animals of Africa

By the same author

He and I and the elephants
No room for wild animals
Serengeti shall not die
Rhinos belong to everybody
Twenty animals, one man
Such agreeable friends
Four-legged Australians

Among Animals of Africa

Bernhard Grzimek

Translated by J. Maxwell Brownjohn

With 46 photographs by the author and 6 by Alan Root

𝔰𝔇 Stein and Day/Publishers/New York

First published in the United States of America
by Stein and Day/*Publishers*/1970

© 1969 *by Verlag Ullstein* GmbH, Frankfurt/M-Berlin
© 1970 in the English translation by Wm. Collins, Sons & Co. Ltd.

Library of Congress Catalog Card No. 78-122418
All rights reserved

Printed in Great Britain

Stein and Day/*Publishers*/7 East 48 Street, New York, N.Y. 10017

SBN 8128-1324-3

Natura enim non imperatur, nisi parendo.

We cannot command Nature except by obeying her.
FRANCIS BACON *(1561-1626)*

Contents

    Foreword *9*
1. How Europe exported apes to Africa *11*
2. The black rhinoceros *38*
3. What the lion's prey feels *67*
4. Do man-eating snakes exist? *77*
5. The ostrich *99*
6. Floating among crocodiles *108*
7. Locusts on the march *124*
8. The hyena changes its image *130*
9. Jumping giraffes *146*
10. Crocodiles that can remember Stanley and Livingstone *155*
11. Animals and fire *178*
12. Escapade in the Sudan *190*
13. The much-maligned gorilla *209*
14. The beast that lived like a man *232*
15. Why don't camels die of thirst? *254*
16. Too few elephants or too many? *261*
17. The Nile, frontier between species *316*
18. The truth about the wild dog *324*
19. Rhino tosses rhino *336*
    Bibliography *353*
    Index *355*

Foreword

We thus meet the artifice, as contemptible as it is brazen, whereby all the natural functions which animals have in common with us and which attest the identity of our nature with theirs, such as eating, drinking, pregnancy, birth, death, corpse, and others besides, are designated by quite different words in their case than in that of man. This is a base trick indeed.

ARTHUR SCHOPENHAUER (1788-1860)

Of the many books which I have written in the past few decades, only four are devoted to Africa. This is the fifth. It describes recent personal experiences in Black Africa and contains sections on individual species found in that continent: elephants, large snakes, crocodiles, gorillas, giraffes, lions, camels, black and white rhinoceroses, ostriches, locusts, hyenas and wild dogs. I have endeavoured to summarize my own experiences together with current information about African fauna accruing from observations and research carried out by others. In so doing, I have become aware that the task is harder and demands more time and space than it would have done only a few years ago.

It was not until 1960, after the Black African countries had gained their independence, that biologists began to devote increasing attention to how the larger African animals live, how they cope with their environment, and what they need for survival. Zoologists can now be found throughout the national parks and whole teams of biologists are at work in the Serengeti Research Institute, just as they are in the newly founded universities of East Africa. In the old days, research was at best limited to pest control, in other words, to the question of how to cull or exterminate species cap-

Foreword

able of harming man's crops, preying on his cattle or transmitting diseases to native inhabitants and domestic animals.

The picture is radically different today. In the colonial era Tanzania possessed only one national park, Serengeti, from which the last British governor-general cut off a large slice only two years before independence. Today, Tanzania has six big new national parks occupying three per cent of the entire country. Annual appropriations have been quintupled and at least three new national parks will be added in the next few years. Serengeti National Park was considerably enlarged by the new president, Dr Julius Nyerere, not long ago, whole villages being resettled for the purpose. Wretchedly poor as it is, Tanzania now devotes a larger proportion of its national income to the conservation of natural resources than the United States. Developments are taking a similar course in the other countries of East Africa. Young African nations are shaming the older countries of the white man, in which many species have become permanently extinct. This is especially humiliating to a citizen of the German Federal Republic, whose politicians have yet to establish a single national park or wild life sanctuary and devote virtually no funds to the conservation of natural resources. Yet our yearning for unspoiled countryside and interest in wild animals grows ever stronger in this overcrowded world of asphalt jungles. In ten or twenty years' time, forward-looking statesmen like Dr Nyerere will be acclaimed as pioneers of a new cultural approach.

No one who sets out to write a book on African wild life today can fail to note that our knowledge has grown by leaps and bounds, particularly in recent years. We are still at the beginning, but we have learned more in this century than in the two centuries that preceded it. I did not fully realize that until I came to write this book.

BERNHARD GRZIMEK
Arusha, the Long Rains, 1969.

1. How Europe exported apes to Africa

A bird that means to fly to Europe will fly there even if you pluck its feathers.
AFRICAN PROVERB

I was quite aware that our project had its dangers.

Most people are only familiar with the droll, lovable young chimpanzees that are trained to appear in variety shows or frolic about in zoos. Once an anthropoid ape of this type has reached sexual maturity at the age of eight or become, at twelve, a fully-developed personality with social demands upon the members of its group, caution is indicated. Because of its shorter legs, a male chimpanzee only comes up to a man's chest when standing erect, but it has muscles twice as powerful, fangs not much inferior to those of a leopard, and a brain more highly developed than that of any living creature on earth with the exception of man. An angry wolf or leopard may sink its teeth into a broom-handle; an angry chimpanzee always bites to draw blood.

The middle finger of my right hand is permanently stiff from the bite of a male chimpanzee. I bear numerous scars from Bambu, another young adult male which I took into my Berlin home with other anthropoid apes when the zoo was bombed. [This episode is described in my book *Such agreeable friends*.] I generally lay flat on my back while Bambu pranced around on top of me in transports of rage, until one day, in desperation, I made myself 'top monkey' with the aid of an oak cudgel. Bambu had to be made to submit because the house might have burned down at any moment.

A zoo director of my acquaintance had his knee-cap half torn out when a male chimpanzee whose cage he used to enter almost daily flew into a fit of rage, and another acquaintance had both thumbs bitten off.

A mere ten days before we embarked on 'Operation Chimpanzee', one of the male chimpanzees attacked Horst Klose, our experienced and athletic keeper of anthropoid apes at Frankfurt Zoo, and escaped. The animal bit through his shoe, severing a toe, and lacerated his thigh and hand. Surgeons at a near-by hospital spent two-and-a-half hours sewing him up, and he remained there for over four months. We finally cornered the bloodthirsty fugitive on the sixth floor of an office building three blocks from the zoo and put it to sleep with an anæsthetic dart. It was all we could do to persuade thirty office workers to leave the premises by a rear exit. They thought it ridiculous to flee from so amusing and friendly a creature.

Such were the animals we planned to release on Rubondo.

I had been introduced to Rubondo eighteen months earlier by Peter Achard, a game warden based in Mwanza at the southern end of Lake Victoria. Achard felt rather forlorn in this corner of Tanzania, formerly Tanganyika and once German East Africa. 'All the tourists and V.I.P.s go to the Ngorongoro Crater and Serengeti,' he told me. 'Nobody comes to Mwanza for a look at my part of the world.'

We boarded a light aircraft and flew the 85 miles to Rubondo. Achard knew what he was doing when he chose Rubondo. I immediately fell in love with the island. It is 24 miles long and, on average, 5 miles wide. Three-quarters of its 135 square miles are covered with forest and the remainder consists of grass-covered hills. Above all, though, Rubondo is uninhabited. The fisherfolk who used to maintain huts and banana plantations on the coast – four hundred of them in all – were resettled on neighbouring islands and the mainland two years previously by their energetic new government. Their banana groves can still be seen at intervals along the island's shores.

Not long ago the new African judge at Mwanza heard a case

Chimpanzees assembled in Europe

involving two men who had illicitly cut wood on Rubondo. 'I realize that the situation is still new to you,' he told them. 'I shall therefore treat you leniently this time.' He sentenced each of them to a year's imprisonment . . . Under the British they would undoubtedly have got off with a fine of a few shillings because all colonial governments tried not to appear 'anti-native' during the closing years of their rule.

So Rubondo now belongs to the animals. Peter Achard had previously captured sixteen rhinos in other threatened areas of Tanzania, most of them badly wounded by poachers or big game hunters. He had talked the Public Works Department, which is responsible for roads and transport, into bringing a car-ferry big enough to hold ten trucks from one of the neighbouring bays of Lake Victoria and transporting the grey giants to Rubondo in their heavy crates. On one occasion the ferry almost sank. Six giraffes made the trip as well, and one of the rhinos has since produced young on the island.

Peter Achard was an energetic and enthusiastic man, and enthusiasm can be infectious even to ageing men like me, who really ought to steer clear of adventuring. As a result, the Frankfurt 'Zoological Society of 1858' circularized all European zoos and on May 16th 1966 dispatched ten large chimpanzees from Antwerp aboard the German African Line's steamship *Eibe Oldendorff*. The chimpanzees remained at sea for over three weeks. The crates in which they were housed were capacious and easy to keep clean. The animals had to travel singly or in pairs because they did not know each other and would soon have come to blows if herded together at close quarters on a voyage lasting several weeks.

They were nearly all young but sexually mature adults and would have commanded two or three hundred pounds apiece on the open market. One pair had been donated by the Copenhagen Zoo, another by the Berggreen Zoo Park of Sweden, and one female jointly by Gottfried Fridh of Stockholm and the Swedish Furuviks Parken. All the rest had been purchased by us or acquired in exchange for other animals. The seven females were full-grown.

Unusual deck cargo

Only one of the males – the muscular Robert – was full-grown, the second being immature and the third little more than a baby.

Because we were anxious that the apes should be hale and hearty when they reached Africa after almost four weeks at sea, our head keeper Gerhard Podolczak had applied for leave in order to accompany them on the freighter. Podolczak had been keeper of the monkey-house for many years, and his wife is currently rearing the first baby gorilla to be born in Germany, a wilful little bundle of joie de vivre called Max, in her own home. We feel entitled to claim some experience in dealing with anthropoid apes at Frankfurt, being the only zoo in the world to have successfully bred all four species of anthropoid apes: chimpanzees, gorillas, orang-outangs, and bonobos. Many chimpanzees and gorillas have been reared as foster-children in my home too.

Captain K. W. Wehlitz took a keen interest in the hirsute passengers on his freighter. He had sturdy wooden sheds built on deck so that the animals need not go below. 6 cwt. of bananas, 12 cwt. of Californian apples, 260 oranges, 3 cwt. of bread, 300 lb. of rice and 18 lb. of tea were taken aboard together with $2\frac{1}{2}$ cwt. of straw and wood-shavings. The chimpanzee villa on deck, 40 feet long and 6 feet wide, was covered with two layers of tarpaulin by the ship's carpenters and even fitted with internal electric lighting.

Travelling at that time of year was no rest-cure, however, either for Gerhard Podolczak or for the chimpanzees. The ship only put in twice, once at Port Said on May 31st in order to load 90 tons of oranges and a second time at Djibouti in order to unload them. Despite air-conditioning and sprinklers, eleven days in the Red Sea proved to be rather more than the monkeys and their keeper could stand.

I myself left for Africa three weeks later, travelling there overnight in the considerably greater comfort of a jet plane. At Arusha (Tanzania), I made preparations for the animals' reception at Mombasa and their journey on. When I took off from Frankfurt the temperature was 32°C. In Arusha it was only 15°C. I quickly bought myself a sweater and an electric heater and began to make

The steamer couldn't wait

phone calls and send telegrams. The train journey from Mombasa to Mwanza would have taken at least a week, so I had dispatched two trucks and two field-cars to the coast. I had also obtained import licences and transit documents for Kenya. It seemed strange that a licence should be required for the importation of African animals into Africa, but there it was. At one time all I need have done was simply send the monkeys to Africa and leave everything to Peter Achard, but poor Peter had – at the age of forty-three – suffered a stroke while on leave in England and had been half-paralysed for months. It was uncertain when or if he would return. A young Canadian Peace Corpsman was deputizing for him, and in a few months one of the Africans to whom we had been giving two-year training courses at our game wardens' school at Moshi was scheduled to take Achard's place at Mwanza. Consequently, I had chosen to fly down there myself.

When the *Eibe Oldendorff* put into Mombasa, 42 other ships were already in the harbour waiting to discharge their cargoes. The freighter would have to wait at least a week, Captain Wehlitz was told, so he restarted his engines and headed for Tanga, the nearest port in Tanzania. Our motorized column followed him along bumpy roads, but Tanga was overcrowded too, so the ship sailed on to Dar es Salaam, and once more the reception committee bumped and jolted along after it. Because the caravan would not now be coming to Arusha, I climbed into my Volkswagen bus and raced to Mwanza in two days, driving via Ngorongoro, through Serengeti, and along the shores of Lake Victoria.

The first Dar es Salaam paper I opened when I reached Mwanza contained a photograph of the chimpanzees accompanied by a ridiculous report alleging that the animals, which hailed from European zoos, were accustomed to nothing but the best Russian tea. It appeared that the chief problem would be how to convert them to drinking plain water in the wild. I don't know which sailor sold this nonsense to the African reporter in Dar, but the same picture and report were reprinted in every German newspaper.

Chimpanzee caravan

Even in the small Lake Victoria port of Mwanza there were African reporters who asked me why we were bringing chimpanzees to Africa from Europe when we could have captured them in West Africa. I explained that nobody captures full-grown anthropoid apes in Africa – they are too dangerous, too strong and quick-witted. Besides, most zoos only want young animals and are not equipped to keep full-grown, sexually mature chimpanzees. They liked to show off the amusing mischievous youngsters and didn't know what to do with them once they had grown up.

I had correctly surmised that we should not have much difficulty in procuring full-grown animals in Europe. It distressed me to see such intelligent and active creatures shut up singly or in pairs in rectangular zoo cages, so I was delighted to be able to give them their freedom. I went on to explain that incarcerating anthropoid apes by themselves was a brutal thing to do – at least as bad as keeping a convict in solitary confinement. All anthropoids were lively and sociable creatures.

What was more, transportation to East Africa from the West Coast, where most chimpanzees are captured, would have entailed greater difficulty and expense than the trip from Antwerp to Mombasa – not, of course, that this had prevented one or two European newspapers from accusing me of carrying coals to Newcastle . . .

The chimpanzee caravan turned up late the same night: Sinclair Dunnett, the young Canadian who had been acting as temporary game warden at Mwanza, my friend Gordon Harvey, formerly game warden at Serengeti, and Ulrich Trappe. All of them were sweat-stained, filthy and exhausted. They had been driving along dirt roads for three days and bivouacking between villages. I knew from personal experience gained on the Ivory Coast what happens if one stops in a village when transporting monkeys or wild animals. Hundreds of locals congregate in the same way as Europeans do when a circus hits town, and the consequent shouting, yelling and teasing drives the animals to distraction.

Lake Victoria is an inland sea

The journey by water to Rubondo is no small undertaking. Lake Victoria is really an inland sea, not a lake. Its 26,800 square miles make it the second-largest lake in the world after the Caspian Sea. Sicily, Sardinia, Corsica and the Canary Islands would fit into it with room to spare, yet there is virtually no traffic and no direct steamer route across the lake. Tourists have yet to discover it.

Taking supplies of food, tents, beds, and five heavy chimpanzee crates, Ulrich Trappe and Dr Fritz Walther travelled from Mwanza to Rubondo, via the southern extremity of the lake, in seven hours. Dr Walther, who had originally begun his study of antelopes at Frankfurt Zoo, was now the world's leading authority on those animals. He had been living at Serengeti for two years in order to study the habits of the 700,000 Thomson's gazelles there. Now he planned to observe how anthropoid apes from Europe acclimatized themselves during the first few weeks on Rubondo.

The rest of us accompanied the remainder of the baggage for another 125 miles, along circuitous and incredibly poor roads, to the lake-side township of Bukindo. From here we made the two-hour crossing to Rubondo, constantly striving to amuse and comfort our swarthy fellow-passengers in their wooden crates.

By evening we were lying in the grass outside our tents near the corrugated iron huts of the African game wardens, devising plans of campaign. The chimpanzee crates stood on the deck of the big boat, off-shore. The sun grew ever redder and more golden, lighting up the fringe of jungle that was Dumacheri, a neighbouring island of much smaller dimensions. We had 135 square miles of the globe to ourselves – and how many people can say that?

What makes Rubondo such a gem of a place is that there are no large predators, no leopards, lions or hyenas. Rare animals could be settled there without risk of their being killed the next day. The island could become a refuge for anthropoid apes, a fact of great potential importance. The demand for chimpanzees for medical purposes would probably become enormous if organs were ever successfully transplanted between man and monkey. Scientists are already working on this problem throughout the world and particularly in the United States, where 30,000 people die of kidney disease alone in a single year. Once the techniques of kidney grafting are developed successfully, it may be only a year before all chimpanzees of the appropriate blood-group are exterminated. Wild chimpanzees now living in Africa total 250,000 at most. Of these, less than 10 per cent belong to the blood group 'O' which would probably be the first to be sacrificed in order to save human lives. It would be a good plan to establish a few island sanctuaries for them here and there.

I have done my best to interest the new African governments in the protection of wild animals on the grounds that large numbers of tourists will visit Africa to see them. Switzerland, Italy, Spain and Yugoslavia are all examples of how much money tourism can bring into a country. My book *Rhinos belong to everybody* describes how I have managed, with the aid of television and illustrated magazines, to attract an increasing flow of visitors to the elephants, zebras, rhinos and lions of East Africa. The new governments have established more national parks and spent more money in consequence – to repeat, more in relation to their national income than the far wealthier United States spends on the conservation of natural beauty and wild life.

However, tourists tend to pay lightning visits to Tanzania from Nairobi and Kenya, just to see the Ngorongoro Crater and Serengeti, and then drive back across the frontier again. Success in settling species of animals which are not found elsewhere in the national parks of East Africa could mean that in five or ten years' time visitors will be boarding ultra-modern steamers and sailing across Lake Victoria to Rubondo. The island could be stocked with gorillas, bongos – even okapis. Situtungas already live there, and the shores of Rubondo still harbour comparatively large numbers of crocodiles, which have been exterminated almost throughout Africa as a result of the fashion for crocodile-skin. Our object, therefore, was to render assistance not only to animals but also to a poor but dynamic new country which does a great deal for its wild life.

If these had not been zoo chimpanzees but animals freshly captured in the jungle, we need only have brought the crates ashore next morning and opened them. They would simply have fled from us and vanished into the primeval forest, never to reappear. Like gorillas, wild chimpanzees are extremely timid and will never attack human beings without provocation. Fifteen years ago my late son Michael and I spent ten days observing chimpanzees living wild in the Nimba Mountains of French Guinea, and we soon discovered how elusive they are. Zoo chimpanzees, by contrast, have no fear of man but are not necessarily tame or friendly. These ones would probably have remained near the huts and established a reign of terror, stealing our food and clothes and possibly doing us bodily harm.

Jane Goodall [who last year married a photographer and has since become Baroness van Lawick-Goodall] spent a total of fifteen months living among chimpanzees in the lonely forests on the shores of Lake Tanganyika. During the first months the animals fled if she came to within 500 yards. Later they allowed her to within 100 yards, after a year to within 30 yards, and in 1964, when she lured them with bananas, to within a foot. After that they stole the clothes off her clothes-line, snatched a half-finished letter from under her very pen, and chased her husband up a hill.

Gerhard Podolczak, who had looked after our chimpanzees on board ship, told us that some of them were very vicious and that they tried to grab and bite anyone they could reach through the bars. In view of the distance that separated us from medical assistance and the nearest hospital, I had no wish for members of my party to risk injuries that might require medical attention.

At dawn, while the lake was still calm, we transported the animals in two batches twenty minutes farther along the coast to a clearing beside the shore where the beach, which consisted of smooth stones, shelved gradually. The heavy crates were carried

The living habits of the wild chimpanzee

ashore on our shoulders and placed in a row close to the water's edge. The animals were given a last dose of Resochin syrup against malaria, a disease to which chimpanzees are susceptible. We had given them prophylactic treatment during the overland journey across Africa. On Rubondo itself the mosquitoes were unlikely to transmit malaria because there was no incidence of the disease there. The risk in towns and villages was far greater.

We up-ended the crates containing the most dangerous inmates so that their sliding hatches could be pulled open from the water with ropes. The ability to swim is as little innate in the three main types of anthropoid apes – gorillas, chimpanzees and orang-outangs – as it is in man, a characteristic which differentiates them from almost all other animal species in the world. It was improbable that they would pursue us into the water, so we should easily be able to evade them by wading in.

A few hundred yards away in either direction stood the deserted banana groves of the original inhabitants. Here, as at many other points on the island, the animals would be able to obtain food until they had learnt to find their own in the forest proper.

We owe our small store of very recently acquired knowledge about the living habits of wild chimpanzees to observations made by Dr Adrian Kortlandt, Jane Goodall in the Gombe Reserve on Lake Tanganyika, and Vernon Reynolds and his wife, who lived among chimpanzees for eight months in the Budongo Forest on the Uganda border. These anthropoid apes roam the forests all day, sometimes covering as much as ten miles. They live on wild fruit, leaves, young shoots, bark, and soft wood. In the Gombe Forest Jane Goodall counted 63 different trees and plants which served them as food. She also saw them kill and devour small monkeys, whereas the chimpanzees in Uganda did not touch meat, even though they toyed with dead dwarf antelopes caught in traps. West African chimpanzees also left smaller animals alone. Gorillas, incidentally, are strict vegetarians. Chimpanzees spend much of their time clambering around in trees in quest of food, but also proceed on all fours along paths which are often of their own making. Michael and I thought we had found some native tracks in the Nimba Mountains until we came upon places where the clearance was reduced to about three feet by twigs and undergrowth. Then we realized what they were.

When groups of chimpanzees meet, they greet each other delightedly. Their fur bristles and they stand erect, placing an arm round each other's shoulders and kissing by thrusting out their lips and waggling their tongues to and fro. They will also take the arm of another animal between their teeth. Thus, chimpanzees broadly resemble human beings in their gestures of friendship.

A large band of, say, 70 chimpanzees will inhabit an area of about 6 square miles of forest which they roam in small groups. These groups are not families. A number of adult males or mothers with small babies will readily join forces. If fruit is ripening on the trees at one particular spot, the whole horde of chimpanzees will often converge with loud cries. Males and young animals quite frequently utter cries while searching for food, whereas groups of mothers are much quieter and seem to head straight for places where others have just found a crop on their wanderings. Baboons are the chimpanzee's only real competitors for food, but Rubondo fortunately had no baboons either.

Chimpanzees like to drum by dancing on resonant surfaces and performing a sort of drum-roll with their hind legs. We found a hollow fallen tree in French Guinea which was used for this purpose. Reynolds observed that the apes also used the frond-like supporting roots of the nyakahimbe or iron-wood tree to drum on. The sound could be heard from almost two miles away.

Even when two large groups of chimpanzees from neighbouring territories meet, no blows or biting result. On the contrary, their normal practice is to clamber up trees, scream and chatter wildly, swing about in the branches, run along them at top speed, jump off, shake small trees and branches, and come to close quarters. Reynolds christened this the 'chimpanzee carnival'. A noisy greeting and introduction of this sort can last for up to an hour. Jane Goodall saw the Gombe chimpanzees performing similar

They kiss like human beings

dances in the rain. The males, fur bristling, grasped a tree-trunk with one paw and ran round and round it with wild abandon while females and youngsters climbed neighbouring trees so as to get a better view of their rain-dance.

But these were all things which it might be possible to observe on Rubondo in years to come. For the moment, we had other worries. We began by releasing those chimpanzees which had proved to be friendly and good-natured on the voyage. Two females, one of them very slender, simply made a bee-line across the clearing and vanished into the forest as though they had always lived there. Others went to neighbouring crates which still contained chimpanzees, stuck their fingers inside, felt the faces of the inmates and exchanged kisses through the bars. (Kissing as a sign of affection is as innate in chimpanzees as it is in human beings.) The half-grown immature male and the smallest chimpanzee of all, still a baby, made straight for us and clung to our legs. The baby climbed up me, kissed me and begged to be cuddled. It ran after us wherever we went, mostly on two legs, which looked quite comical.

That, however, was precisely what we did not want. The young were to remain with the adults – not a difficult thing to ensure under normal circumstances. All chimpanzees are well disposed towards the young. Even adult males will often play with them for half an hour on end, allowing them to do gymnastics on their backs, tease them and pounce on them. We duly took the little fellow to two females which were sitting at the edge of the forest. It went to them, whereupon one female put her arms round it and embraced it. However, the youngster soon returned to us human beings. We did not venture to shoo it away in case it became angry and started to cry, which might have prompted its elders to defend it and bite us.

Cautiously, from out in the water, we opened the crate containing a large female noted for her spiteful disposition. This had a heavy, barred door, not a sliding hatch, and it was hard to see how she had ever been lured inside. Zoo staff generally put a travelling crate at the mouth of a cage with sliding hatch raised, chase or lure

Reluctant to be free

the animal inside, and then slide the hatch shut. A crate with a hinged door has to be pulled back to allow the door to close, and that is the moment when an animal can escape.

The difficult female showed absolutely no inclination to emerge. She merely opened the door a few inches and peered out. As soon as the baby approached, she took it into the crate with her.

We grew impatient, but the big female took no notice. When she eventually emerged after a good fifteen minutes, she held on to the barred door with one paw and surveyed her immediate surroundings with extreme care, obviously reluctant to leave her place of refuge. She let go of the barred door for a moment and it slammed shut. Startled, she turned back and opened it again. Being keen to get her off our hands and load the empty crates on to the boat, we splashed her with water from a safe distance.

Zoos not infrequently encounter creatures that, having spent days or weeks in their travelling quarters, feel safer inside and are reluctant to exchange them for unfamiliar surroundings.

Late that afternoon we returned by motor-boat. All the chimpanzees had vanished. The piles of bananas, apples and bread which we had left behind lay there untouched. Studies conducted in the Budongo Forest indicate that chimpanzees can live for weeks on leaves and shoots alone, even when no fruit is obtainable, but this would be a new and unfamiliar diet for our zoo specimens. When we came ashore the large male emerged from the forest, followed by the two young males, and greeted us with a handshake. There was no sign of the females, and we wondered if the little males would link up with them. Male chimpanzees are not naturally jealous of each other. Jane Goodall once watched a female copulate with seven different males in succession.

Robert, the big black male, gradually became more and more agitated. His fur stood on end, making him look twice as bulky, and he started to walk on two legs. We hastily retreated into the water – in our shoes and stockings – and back to the boat.

It was July 23rd, and all our anthropoid apes were at last installed in their new home. They had been confined to their crates

Exploring the island

for at least five weeks – some of them even longer, counting the overland journey to Antwerp. Presumably, they felt just as relieved as we did.

Next morning we trekked across the peninsula which separated our camp from the place where we had left the chimpanzees. We had to scale one or two hills whose surface consisted entirely of rock fragments, but the ankle-breaking jumble of stones was overgrown with chest-high grass as yellow as ripe corn. Bush fires had burnt the grass everywhere on the mainland, but not here. The hills were higher and harder to climb than they appeared from lake-level. Between them lay valleys containing jungle, marshy ground, reeds and sruhes taller than a man. We followed paths that had been worn by rhinos. They were easy to walk on, and interspersed with resting-places flattened by the creatures' bulky bodies. Or were they hippos?

It was probable that here, as in many parts of Africa, the hippos only came ashore at night, but who could rely on that? I did not relish the idea of coming between a frightened hippopotamus and

Page 25:
We chose a fairly level spot on the shores of Rubondo to unload the crates containing the chimpanzees. Although they had to be manhandled ashore, this gave us a chance to take refuge in the water after opening them. I must have caught my bilharzia on this occasion. Some of the adult zoo chimpanzees were anything but harmless. One of them had already put a keeper in hospital. Another particularly spiteful male caused considerable mischief on Rubondo eighteen months later. Sinclair Dunnett is here seen opening a travelling-crate with great care. He could retreat into the water at a moment's notice because anthropoid apes are non-swimmers.

Page 26:
A chimpanzee which becomes angry or agitated and wants to impress another stands erect and hunches its shoulders. Its fur also stands erect, particularly on the back and arms. Rage and excitement have the same effect on us, except that we only get gooseflesh because we have no impressive growth of hair on our shoulders. This is one of the chimpanzees released on the island of Rubondo, Lake Victoria.

Page 27:
The full-grown chimpanzees released on Rubondo quickly became acclimatized and have already multiplied. The island is very suitable for stocking with rare species of wild life because no large predators are yet to be found there. According to recent research by Dr Adrian Kortlandt, chimpanzees deliberately strip branches of leaves and use them as weapons.

The car-tyre that was a python

its watery place of refuge. A few years ago, while I was visiting Queen Elizabeth National Park in Uganda, an African girl had met her end in just such a manner. We duly conversed in loud voices and banged our walking-sticks against tree-trunks so that any hippopotamus or rhinoceros would hear us from far enough away not to be taken by surprise.

Just as we were traversing an expanse of yellow grass, we heard the thudding footsteps of a heavy animal circling us at a gallop. It was amusing to note how hastily we all headed for the nearest clump of trees. In fact, an African fisherman had been killed by a female rhinoceros three months earlier, not far from there. The fisherman, who had landed on the island illegally, came across a sleeping female and her calf and pelted her with stones to rouse her. He roused her so successfully that she killed him on the spot.

We descended a slope into the forest. The scree was covered with dense swathes of long dry grass, bent double. Suddenly, lying in front of me in the midst of the matted yellow grass, I caught sight of a car tyre. For a fraction of a second I felt annoyed that anyone should have dumped such rubbish in the wilderness – then I realized that I was looking at the coils of a fat reticulated python. I just managed to grab the snake's tail, but its purchase on the grass and underlying scree was so strong that it easily broke my grip. Pythons are not poisonous, as everyone knows. It dawned on Sinclair Dunnett and me at the same moment, that although each of us had brought snake serum and syringes to Rubondo, we had – as usual – left them behind in camp. One sees relatively few poisonous snakes in Africa and rarely gets bitten, but if the worst had happened on this occasion we should have had to walk or be carried for two hours in order to reach the serum.

Page 28:
135 square miles of island at the southern end of mighty Lake Victoria have been cleared of human inhabitants and placed under a strict conservation order. In 1966, apart from guereza monkeys and roan antelopes, we stocked the island with ten chimpanzees from European zoos. They became well acclimatized and have since produced young.

Never argue with a chimpanzee!

I calculated that the spot where we had released the chimpanzees lay beyond the next hill, but no sooner had we entered the intervening strip of forest than we suddenly came across two large female chimpanzees, one of them having the dangerous disposition I mentioned. They rose from a seated position with their fur bristling, always a sign of excitement and frequently one of hostile intent. Then they came towards us. There was no point in running away, nor had we any means of fighting them off. The larger of the two females put her arm round my shoulder and opened her mouth, baring a set of large and powerful fangs. She took my calf between her jaws, then my entire forearm. Finally she kissed me. So far, all was sweetness and light.

Gerhard Podolczak and I, who were familiar with chimpanzees and knew how readily excitement can turn into hostility, looked round. We hurriedly made our way to the open clearing, and – sure enough – the females preferred to stay in the forest rather than follow us out into the sunlight. On the other hand, one of them removed Dr Walther's binoculars. He wisely offered no resistance and allowed her to pull the carrying-strap over his head. Meanwhile, the big male known as Robert had joined the party. The three chimpanzees expertly opened the leather case and took out the binoculars. They examined them closely and even peered through them, but from the wrong end. Dr Walther tried to swap his field glasses for a handkerchief and a bag, but this aroused more resentment than approval. The animals merely appropriated his handkerchief as well.

It is improbable that any other animal – horse, rhinoceros or antelope – would spend so much time contemplating an object as useless as a pair of binoculars. But anthropoid apes are inquisitive. They can handle implements as human beings do. They will, for instance, use a stick to bring something within reach of their cage; they convert two short bamboo canes into one long one by trimming the thinner with their teeth and inserting it into the thicker; they moisten leaves, plunge them deep into termites' nests or ant-heaps, withdraw them and use their lips to remove the insects that adhere to them. They chew up leaves and suck them dry, thrust the masticated wad into hollow tree-trunks containing water which would otherwise be inaccessible to them and soak it up, then convey the home-made sponge to their mouth. Chimpanzees wipe themselves with leaves when soiled with earth, blood or excrement. Mothers do likewise on the rare occasions when their babies dirty them. Although they never throw stones with deliberate aim, chimpanzees do fashion clubs out of branches and use them to pelt or strike adversaries such as leopards.

Cautiously, we began to edge into the forest in the opposite direction while the three chimpanzees were preoccupied with the binoculars. Robert spotted the move and came after us. He climbed up Dr Walther and took his chin between two rows of teeth, which included powerful canines. Menacing as it looked, this was a friendly gesture. But the animal's fur was standing on end. Robert now began to stamp the ground, his fur growing ever bushier. He raced between Dr Walther's legs at top speed, up-ending him. Sinclair Dunnett suffered the same indignity. Then Robert grasped his hand and began to circle him as if he were a May-pole, screeching and bristling.

'Make for the undergrowth, quickly!' I called. Sinclair did so. He plunged into the trees where Robert could no longer perform his whirlwind gyrations. We pressed on into the forest so as to skirt the chimpanzees' clearing in a wide arc, but Robert did not stay behind with the females. He accompanied us, still bristling. After a while he snatched Sinclair's panga out of his hand. I approached the male chimpanzee with my open hand extended palm-upwards in the begging gesture which is habitual to chimpanzees as well as men. Robert surrendered the big bush-knife without demur but promptly took possession of my walking-stick instead. He was growing more and more excited. I tried to visualize how we would transport an injured man all the way back to camp.

I was duly grateful when Robert took Sinclair Dunnett by the hand and followed him on three legs. He not only declined to release Sinclair's hand but hung on to it so hard that Sinclair was

Escape by water

compelled to half-drag him along. Robert also refused to allow Sinclair to take the initiative and lead him with the left hand.

Our position was becoming rather awkward. Chimpanzees are venturesome creatures and exceptionally fond of wandering. It was quite conceivable that Robert would stay with us for hours – even accompany us as far as the camp. Thanks to the chimpanzees' system of communication by calls and drumming, we might soon have the rest of the colony on our hands. How could we shake Robert off? Go back to the females? They might attack us or tag along too. We were all very agitated and conversed in half-whispers, although there was really no need. Then we hit on our old refuge: water.

We promptly marched downhill towards the lake-shore, through tall and untended banana-trees, and walked straight into the water in our shoes, stockings and shorts. Crocodiles? We hoped they would steer clear of four men. Like all the islands around, Rubondo's shores were fringed with dead trees denuded of bark. For the first time in eight decades, the level of the lake had risen some six feet because of heavy rains in the previous three years. Many old trees had met their end as a result. I quickly stuffed my wallet and the contents of my trouser pockets in the breast-pocket of my khaki shirt. Then, waist-deep in water, we waded along parallel with the shore, probing for hidden holes with our sticks and trying to avoid patches of mud. It was best not to dwell on the thought of bilharzia, the small worms which take up residence inside the human body. (A year later the doctor in Arusha discovered that I had, in fact, picked some up.)

The big male chimpanzee followed us along the shore. At a spot where the reeds were dense he climbed an old tree to get a better view of our progress. We wondered how long he would keep it up. The shore was overgrown with rushes, but it was still easier for him to negotiate them than for us to wade through the lake. Another fifteen minutes and we had lost sight of him. We continued to converse in whispers and sign-language for fear of attracting his attention. Our hopes gradually rose, but our route was barred by the thorny top of a fallen tree. Being unable to skirt it on the lake side because the water was too deep, we had to crawl ashore on all fours through a veritable maze of creeper, fortunately of the thornless variety. One after another we bored a tunnel through it and wormed our way for fifty or sixty yards until we eventually came to a stretch of grass and bushes. We looked round warily, but there was no sign of Robert or the rest of the colony.

Then I found that the wallet containing the money for our trip had fallen out of my shirt pocket, which I had omitted to button in all the excitement. It had probably slipped out the first time I bent down in the undergrowth. Dr Walther and I crawled back the whole way on our bellies. Sure enough, the wallet was lying deeply bedded in twigs and creeper at the very spot where we had entered the thicket.

We marched back up hill and down dale in our wet shoes and trousers. We were dry by the time we arrived but our legs and ankles were torn and scratched, not that we had noticed at the time. My shirt was in rags but my trouser-creases had survived our under-water march – a tribute to the wonders of modern textile chemistry.

When we returned, much more comfortably, by motor-boat the same afternoon, not a monkey was to be seen. The bananas were still lying there untouched. We found the binoculars case on the edge of the forest, its leather torn and chewed to pieces. The binoculars themselves came to light after a long search among the trees – broken, needless to say. We saw no sign of any nests within a wide radius.

The presence of chimpanzees in a forest is most readily betrayed by their nests. These they construct every evening between five p.m. and seven, when darkness falls, by snapping and interlacing branches and topping them with broken twigs. Nests vary in altitude from thirty to as much as a hundred feet. Unlike gorillas, chimpanzees never deposit faeces in such nests. They generally excrete during their wanderings while seated on a broken branch

The temperamental engine

so that the stool falls to the ground. They have also been seen to complete the performance by wiping their hind quarters with leaves.

After a few days we left Sinclair Dunnett behind on Rubondo to keep an eye on the chimpanzees and supervise the African game wardens. To help them prevent fishermen slipping across from neighbouring islands to steal fruit from the neglected banana plantations, we purchased a heavy wooden boat for them, powered by an outboard and large enough to hold twenty people. Sinclair Dunnett was soon to be succeeded by the young German gamekeeper Ulrich Kade, who has now been stationed on Rubondo for two years. So far, the experiment has gone well.

Feeling relieved, I boarded the big inboard motor-boat and headed straight for Mwanza. The diesel engine puttered along for three hours and then cut out. The captain and the engineer worked on it and got it going again. This time it ran for less than an hour before relapsing into silence. It was even longer this time before the black smoke started spewing sideways from the hull. Islands glided past us, cormorants watched us as they perched on bare branches by the shore. Beneath the trees on small islets stood reed huts cruder and more primitive than almost any to be found elsewhere in Africa. Next time the engine ran for only twenty minutes, then ten, then five. I checked its progress by my watch, growing more and more uneasy. Would it refuse to start again altogether? The waves were becoming higher and capped with foam. It was the dry season, fortunately, so there was no fear of a storm – no fear, either, that the stoutly constructed boat would founder. Nevertheless, if the engine broke down completely the wind would inevitably carry us far out into the open lake. We should have to spend at least one night afloat, even assuming someone started a search and came to our rescue. All kinds of possibilities rear their ugly heads on such occasions.

The waves were now breaking over the boat and drenching us to the skin. The native crew cut up old sacks and tied them between the roof-stays on the weather side, but the waves broke over

Lions and film stars

them just the same. I opened my bag and brought out an umbrella which my companions had previously laughed at. The three of us crouched behind it for five long hours.

We reached Mwanza that night, after nineteen repairs and fourteen hours afloat. Next day I packed Head Keeper Podolczak into my Volkswagen and took him to Serengeti, driving for three full hours through tens and hundreds of thousands of gnus, zebras, Thomson's gazelles. In the evening, at Seronera, the national park headquarters, I was able to point out half a dozen lions for Podolczak's benefit and show him a leopard in a tree. The following day we drove through the mountains and down to Arusha to catch the plane. A single night's flying brought us to Frankfurt. As luck would have it, the German film star Hardy Krüger was sitting between us during the flight, so it was very late before we stopped swapping stories and got some sleep.

For the first time, anthropoid apes from Europe had travelled back to their original home in Africa. Good luck to the chimpanzees of Rubondo!

*

Twelve giraffes had already been installed on Rubondo between July and November 1965. A group which included two young was observed in 1967, and several more groups were sighted in 1968 and 1969. Sixteen full-grown rhinos were brought to the island during 1964-5. Several sightings were made during the years that followed, including two of young, one in 1966 and another in 1967. Flying over the island in September 1966, the Serengeti game warden Myles Turner saw four head of cattle which behaved like buffalo, and in 1967 a sighting was made of three wild goats which must also have owed their existence to earlier settlers on the island. In October 1967 two male and two female roan antelopes were released, two of which were seen again in February 1968. I also saw their tracks in February 1969. In June and July 1967 we settled a total of twenty guereza monkeys on the island. Some of them I saw from the boat in February 1969, in the trees along the shore. I also managed

Chimpanzee reinforcements

to film some giraffes in the interior of the island while flying over it the same day.

Of the ten chimpanzees, three male and seven female, whose release I have just described, only one was a full-grown male. The two young males did not, unfortunately, link up with the adult chimpanzees but tended to keep to themselves. They readily approached human beings in the first few months but fled from an unknown boat six months later. It took them a good year to lose their tameness. We thought it too risky to have only one adult and sexually mature male in the group, and so, in 1966, two adult males from zoos were sent separately from Europe to Rubondo by air. One of them was Jimmy, the one-eyed chimpanzee which had seriously injured Keeper Klose during its temporary sojourn at Frankfurt Zoo. In September 1967, or fifteen months after being released, the male chimpanzee Robert turned up at the game wardens' new second camp with two pregnant females. The animals tore up sacks of sugar and grain, scattered things about and behaved in a thoroughly refractory manner. Ulrich Kade was bitten in the hand while trying to drive them away. The game wardens repelled later invasions of the camp by firing shots in the air. The chimpanzees did not reappear after that, but in February 1968 two females were observed with young.

In November 1967 one of the island's game wardens was shot by poachers. The culprit was recognized, although he escaped, and the police succeeded in arresting him at his native village on the mainland two days later. He was hanged in 1968. As a rule, the courts passed sentences of six weeks' imprisonment for unauthorized landings on Rubondo and six months' imprisonment for attempted poaching. A fishing-boat which put in at the island without permission in March 1968 was attacked and capsized by a hippopotamus. One of the crew was drowned.

On October 10th 1968 one-eyed Jimmy broke into the new camp and immediately attacked Game Warden Lucas Seremunda. Determined to keep the animal away from his children, the man refrained from running into his house and made for the lake-shore, but Jimmy caught him before he reached the water. In a rage, the chimpanzee bit him in both hands and would not let go until hit on the back with a stick by another game warden. Then it ran off. Six days later it reappeared and attacked Game Warden Daniel Obaha, who was sitting outside the camp reading a book with his rifle at his side. Obaha tried to run inside a hut when he saw Jimmy heading for him, but the furious animal tore a piece of flesh off his leg and forced its way into the hut as well. It then closed the door from the inside. In the fierce struggle which followed Obaha lost the little finger of one hand and had the other badly mauled. He fortunately managed to wrench the door open. One of the porters picked up Obaha's rifle, which was lying on the ground, and shot the chimpanzee dead. However, the rest of the chimpanzees and the other animals on Rubondo are obviously thriving.

2. The black rhinoceros

The elephant's and rhino's concrete skin
hides much that's soft and mellow,
as all who really peer their jaws within
will tell you.
They're shot by soft-skinned men with hard interiors.
I think you go to heaven, you great chimeras!

JOACHIM RINGELNATZ

To face a black rhinoceros on foot, not seated in a car in the now customary way, is a diminishing and humbling experience. It instantly recalls the numerous stories of fierce and lethal onslaughts which one has read in books about Africa. A hook-lipped rhinoceros can weigh as much as two tons, measure 12 feet from the tip of its nose to the base of its tail, and has a shoulder-height of 59-63 inches. It is thus one of the world's largest land animals, small though it seems in comparison with many of the extinct members of its family, e.g. the Indricothericum asiaticum, the largest mammal yet identified. The latter was 16 feet tall, or the height of a giraffe, and 23 feet long. The bones of this gigantic creature, approximately 35 million years old, have been found on the banks of the River Chulka in Kazakhstan. Rhinos quite similar to those alive today were still distributed throughout Europe and Asia in comparatively recent times. In Cracow Museum I saw a stuffed woolly rhinoceros only ten thousand years old which had been found well-preserved in oil-bearing and saliferous strata at Starunia in 1929.

The black or hook-lipped rhinoceros is as little black as the white or square-lipped rhinoceros is white. Depending on the area in which it lives and the mud or dust in which it rolls, its fundamentally slate-grey colouring can be so effectively concealed that it appears sometimes white, sometimes reddish, and – in volcanic regions – entirely black. It has no body-hair except on the tip of its tail and edges of its ears. It also has no sweat-glands, hence its predilection for mud-baths. The rib-like folds along its flanks are wholly independent of the actual ribs beneath. The animals have no incisors or canines, merely seven molars in each half of the jaw. At the Frankfurt Zoo we found discarded teeth in the dung of a bull rhino aged between eight and nine, the stage at which the animals may be regarded as full-grown. Towards the end of the last century, museum zoologists tended, on the basis of individual horns and bones and of the few African rhinos kept in zoos, to recognize a whole series of species and subspecies of hook-lipped rhinos. Today, all hook-lipped rhinos are assigned to the species Diceros bicornis, the only possible exception being the Diceros bicornis somaliensis found in the extreme north of Kenya and Somaliland, which is some 10 per cent lighter.

The most impressive feature of this rhinoceros is its two horns. Visitors seeing our specimen at Frankfurt for the first time subconsciously feel them between their ribs, yet zoo rhinos very seldom attain the remarkable length of horn which many wild hook-lipped rhinos carry about with them. The longest horn ever found (on a rhino in Kenya) measured 53 inches along the outer curve. 'Gertie' one of the two cow rhinos at Amboseli with unusually horizontal and upward-curving front horns, later gained the world record with 54.3 inches. For many years she was the most-photographed wild animal in the world. The cow 'Gladys', who lived in the same place, had a monstrously developed horn of the same calibre. In 1955 she broke $17\frac{1}{2}$ inches off, but photographs disclose that both these animals had put on approximately $17\frac{1}{2}$ inches more horn in six to seven years.

Rhino horns do not repose on a knob of bone like the horns of an ibex or cow. They consist to some extent of hairs baked together – in fact a rhino horn can actually 'fray'. Only slight bleeding results when a rhino's horn snaps off, as it does occasionally, and the horn grows again. The front horn is invariably longer than the rear one, though it is said that rhinos with horns of equal length used to exist in one or two regions of Africa where the creatures have since been exterminated.

Powdered rhinoceros horn is sold in Far Eastern apothecaries' shops, particularly in China, as an aid to stimulating sexual appetite in man. This is why rhinos, which are easy to kill, have suffered so persistently at the hands of poachers. Years ago, people were paying as much as £3 per lb. of horn on the black market. Its value is based on the same sort of superstition as that which enables Chinese apothecaries to sell dragons' teeth and other weird things for medicinal purposes. John A. Hunter, whose sad boast it was to have shot more rhinos in his lifetime than anyone else, grated some rhino-horn and brewed it into a dark-brown tea. Although he drank several doses of the brew he felt no effect of any kind, possibly because he had no faith in it or possibly – he wrote – because he lacked the right stimulus. The medicinal properties of rhinoceros horn have recently been subjected to exhaustive tests at the instigation of Dr A. Schaurte, with equally negative results. Needless to say, the demand for it persists. Superstition has been responsible for exterminating many species in the course of time.

Triple-horned rhinos have sometimes been encountered, especially in the vicinity of Lake Young in Zambia (Northern Rhodesia). There has even been a five-horned hook-lipped rhinoceros, and others have horns growing out of their bodies. Many people have smiled at the armoured rhinoceros with a small shoulder-horn which figures in the celebrated Dürer engraving and has so often been copied by other artists, but it may possibly have been based on a live specimen.

I myself have photographed earless rhinos, most of which have a truncated tail as well, not only in the Amboseli district of Kenya but also in the Ngorongoro Crater. It is a favourite story among the Masai that cutting the ears off sleeping rhinos is one of their tests of courage, but some of the animals are actually born without ears.

How fast can a rhino travel?

Having observed the earless Pixie at very close quarters, I gained the impression that this animal can narrow – indeed, seal off – the entrances to its auditory canals at will. The only way of settling the question, I suppose, would be to examine the ear-muscles of a dead rhinoceros. Pixie, incidentally, was born in 1953 to Gertie, whose ears are well-developed.

Black rhinos are remarkably shortsighted. It is evident that they cannot distinguish a man from a tree-trunk, even at a range of 40 or sometimes 20 yards, a fact which accounts for much of their behaviour. Their hearing is considerably more acute, and their paper-bag-shaped ears swivel rapidly in the direction of unfamiliar sounds. Their sense of smell is excellent and by no means inferior to a dog's. They follow the tracks of other rhinos by scent. Mothers and young which have become separated may be well within visual range from a human observer's point of view. Far from going up to one another, however, they sniff their way along the ground until they pick up and follow the appropriate scent.

I have been charged by rhinos several times while sitting in a car but only once on foot, when I was in the Ngorongoro Crater. On the last occasion I managed to dodge round a field-car and dive into it, moving faster than I would have believed possible. Meinertzhagen clocked between 50 and 56 kilometres per hour (over 30 m.p.h.) on the speedometer of his car while rhinos were charging him, and on another occasion a rhino chased a moving car for four hundred yards at a speed of 28 m.p.h. Normally, however, the animals travel very slowly even when not grazing, certainly nowhere near as fast as a man walking. I have never met anyone who saw a rhino swimming properly through a river or deep lake, although the animals are passionately fond of wallowing in the shallows and grazing on rushes. They *can* swim, of course. When an attempt was being made, during the damming of the artificial Kariba Lake in Zambia, to rescue animals which had taken refuge on the gradually dwindling islands, one rhinoceros attacked a boat and ventured out of its depth into surprisingly deep water. It almost vanished beneath the surface with only its nose, ears and eyes protruding above water-level. A very slight swell would have been enough to submerge the animal repeatedly.

Hook-lipped rhinos may look ungainly, but they climb quite high into the mountains and have been found at altitudes of 8,800-9,500 feet in the mountains of East Africa. Black rhinos live in dense bush, sparse forest, open grassy plains, even in areas of semi-desert. The one thing they dislike is heat combined with humidity, which explains why they have never penetrated the rain-forests of the Congo Basin or the jungle of West Africa. They were not found throughout Africa even in earlier times. Their habitat was probably restricted to the southern tip of the continent – the environs of what is now Port Elizabeth, Transvaal; the southern part of Angola and from there up to the West Coast; the whole of East Africa, Mozambique, Tanzania, Kenya, Somalia, and as far north as Ethiopia; and a strip extending between the Sahara, the Congo and the jungles of Nigeria, down to the region of Lake Tchad and the French Cameroons. Hook-lipped rhinos did not occur in many areas of the great plains of Central East Africa, only two fingers of which reach out towards West Africa, e.g. along the Kenyan and Tanzanian coast or between the Zambezi and Chobe rivers. Since the penetration of Africa by Europeans, they have become extinct over wide areas, e.g. south of the Zambezi. It was not until 1930, by which time they had almost vanished from the French African colonies, that a few were preserved by the introduction of strict protective measures. The total number of black rhinos surviving in Africa today is estimated at 11,000-13,500, of which a large proportion (3,000-4,000) live in Tanzania.

It is almost impossible for us to grasp, even today, the slaughter perpetrated on the hook-lipped rhinoceros by hunters, most of them white. No fewer than eight hundred rhino-horns were exported from the sultanate of Fort Archambault, in the Lake Tchad area, during 1927. A professional big game hunter named Cannon shot approximately 350 rhinos in less than four years. He and another big-time butcher named Tiran were particularly active in the Cameroons, Ubangi and Tchad. They periodically transferred their attentions from elephants to rhinos, which were easier to kill than elephants and possessed horns which were appreciating in value. The natives, equipped by these men with modern weapons, took an active part in the slaughter or operated on their own account. The British big game hunter John A. Hunter boasted of having dispatched over 1,600 rhinos in addition to more than 1,000 elephants. This he did only partly on behalf of the government, which wanted, for instance, to clear Wakambaland in Kenya in preparation for the arrival of settlers. Hunter shot 300 rhinos there in 1947 and a further 500 in the following year. It turned out later that the area was scarcely suitable for human habitation. Even harder to comprehend are the so-called sportsmen who used to travel round Africa shooting as many unsuspecting creatures as possible for the mere fun of it and without financial reward. A Dr Kolb is reported to have mowed down more than 150 rhinos in East Africa.

Psychologists ought to make a special study of the mentality of such people, based on their letters and accounts of their activities. They obviously differ completely from the European hunters who conserve wild life, rent shoots and pay compensation for damage done to crops by game so as to maintain stocks and improve them. It may be assumed that a sense of personal inferiority, suppressed destructive urges and a certain thirst for glory was latent in such people, because big game hunting was always painted in particularly heroic colours back home. The celebrated British explorer Frederick Selous (1851-1917), who was educated in Germany, never in all his long years in Africa heard of a case in which a European rhinoceros-hunter had been killed by a rhinoceros.

Today, the black rhinoceros is practically extinct in South Africa, except for a few in protected areas. There are only one or two in Rhodesia and Malawi and slightly more in Zambia, especially in the region of the Luangwa River. The few rhinos which used to inhabit the Southern Sudan probably disappeared thanks to the war and the recent spread of fire-arms. Stocks in Mozambique are put at roughly 500 head, in Angola at 150, in South-West Africa at 280. Were it not for the establishment of national

They live to schedule

parks and similar protected areas in the last few decades, the fate of the black rhinoceros in Africa would no doubt have been sealed long ago.

Unlike elephants, which enjoy wandering, rhinos very seldom re-establish themselves voluntarily in areas where they have once been exterminated. It is not too difficult, of course, to resettle them artificially by capturing them elsewhere and transporting them to a predetermined area. This was done at the Garamba National Park in Ruanda during the 1950s, and in recent years – as I have already mentioned – we captured sixteen rhinos, most of them badly wounded, in the game areas of Tanzania and ferried them to the island of Rubondo in Lake Victoria. It is, however, a natural characteristic of the hook-lipped rhinoceros to cling stubbornly to its home territory in the face of human settlers and growing disturbance.

Our knowledge of the living habits of these grey giants has increased in recent years, now that it derives from zoologists and game wardens patiently at work in national parks rather than from big game hunters. In contrast to many other species of animals, black rhinos have no fixed territories from which they expel other members of their species. On the other hand, the same animal can be found engaged in the same activity at the same spot, at least at the same time of year and often at precisely the same time of day. Once a day, for instance, a rhino goes down to drink along a broad and well-worn track. The distance between feeding-

Page 45:
Lions used to live in Europe too. They probably retreated southwards as forest gradually gained a hold on their hunting grounds, but still existed in Greece at the time of Odysseus and Homer. The lion is an infrequent feeder. It can devour fifty or sixty pounds of meat at a single meal.

Pages 46-47:
In the Etosha Pan in the huge Etosha National Park in the north of South-West Africa, unusual subsoil conditions ensure that some water-holes retain water throughout the year, even during the extremely long dry season. Car-borne visitors can always watch groups of animals drinking there in relays, e.g. these Chapman's zebras and greater kudus. Kudus are one of the commonest species in this area.

The rhinos of Ngorongoro

grounds and watering-place may be five miles or more. The animals usually start to graze at noon and spend the rest of the day in the shade of a tree or in a mud-bath. At night, down at the pool, they indulge in high-spirited games, chasing each other with growls and snorts. In areas where they are not hunted, e.g. the Ngorongoro Crater or the Amboseli district, they can be seen on completely open ground all day long. Because of their passion for mud-baths they sometimes become inextricably stuck or are gnawed by hyenas while helpless to defend themselves.

It used to be thought that black rhinos lived in fixed areas. That this is not the case has emerged from increasingly detailed observations made in the Ngorongoro Crater, Tanzania, which covers an area of 100 square miles. My son and I counted 19 rhinos from the air in January 1958. Molloy's score in March 1959 was 42. Hans Klingel, who kept watch on the Ngorongoro rhinos from June 1963 to May 1965, identified 61 of which 34 could usually be seen the whole time or for periods of several months. These animals seemed to be more or less permanent residents of the crater floor. The largest number of different rhinos sighted from the air in a single day was 27 on February 18th 1965 (Turner and Watson); the lowest figure, 10 on October 8th 1963. J. Goddard, a biologist who lived in the crater for three years up to and including 1966, constantly photographing and familiarizing himself with each individual animal, sighted 109 different rhinos in the crater during this period. He assumed that the vast majority of these animals lived in the area above the lip of the crater throughout the year. In similar

Page 48:
A lion on the attack. This one had separated from the rest of the pride with the lioness in the background, which was on heat, and greatly resented being disturbed. It angrily charged my field-car again and again but did not spring at it in earnest. (Etosha National Park.)

Animals at watering places often take fright and race off wildly for a short distance, like these springboks and gnus at Etosha National Park, South-West Africa, whose 22,000 square miles make it the largest game reserve in the world. The first to designate it a protected area was the German governor, von Lindequist (1907). The heart of the park is the Etosha Pan, which is over 80 miles long and devoid of water for most of the year.

fashion he identified 70 different rhinos in the vicinity of Olduvai Gorge, the celebrated site in the Serengeti Plain where remains of early man were discovered. Klingel repeatedly encountered most of the permanent inmates of the crater – particularly bulls – in certain specific areas. Many of these were never seen outside. It can nonetheless happen that individual animals of both sexes leave their regular haunts and settle permanently in another area.

Black rhinos have a particular fondness for twigs, which their pointed upper lip can grasp like a finger or hand. Even when one sees them grazing on a grassy plain, they may really be uprooting tiny little new bushes. Fraser-Darling saw one rhinoceros uproot and devour 250 little acacia plants in a single day. How greatly rhinos can transform the African landscape in this way, and what results their extinction may have in many areas! An observer in Natal, South Africa, saw two black rhinos felling a comparatively strong mtomboti-tree (Spirostachys africanus). One of the animals gripped the trunk between its two horns and exerted pressure by shifting the whole weight of its body in a circle. The tree snapped and fell. When it was on the ground, both animals began to browse on the tips of the young branches.

Another rhinoceros was standing on the extreme edge of a small cliff, browsing on twigs. In attempting to get at a twig which was almost out of reach it must have rested too much of its weight on the overhanging rock, which broke off. The animal fell more than 30 feet and was killed. Black rhinos eat the spiky twigs of the thorn-bush and are not averse to the sticky white sap of the euphorbia.

They occasionally ingest their own dung, in the wild as well as in captivity. Klingel spent several days watching a group of four animals repeatedly eating gnu-dung. During this period, several hundred gnus were grazing in an area which had recently suffered a bush-fire and was covered with short, very fresh grass not exceeding 3 or 4 inches in height. The rhinos consumed dung which was either fresh or only superficially dry. They picked up a large accumulation and chewed it, dropping fragments in the process

The rhino's needle

but swallowing the bulk of it. They touched no plants while engaged in this form of activity, but went straight from one heap of dung to the next. It is probable that they were making good a deficiency of minerals or hormones in this manner.

Black rhinos often use their horns to dig up earth containing mineral salts. Although they are also said to use them to scatter their own heaps of dung, they normally do this with their hindlegs like a dog strewing freshly deposited dung with earth. The rhinoceros scatters its dung about in a similar fashion. According to a Zambian legend, the rhinoceros roots around in its own dung with its horn in the hope of finding a needle which it has swallowed by mistake. When God created the animal world, he gave the rhinoceros a needle to enable it to sew its hide to its body. According to another legend, the rhinoceros scatters its dung at the behest of the elephant, which dislikes other animals to leave dung-heaps as large as its own.

During 1964 and 1965, the Swiss zoologist Rudolf Schenkel spent several months studying the black rhinos at Tsavo National Park in Kenya. He ascertained that they do not excrete faeces and urine simultaneously, like elephants, but that different rhinos, both male and female, may deposit dung on the same heap. In rare cases, dung is also deposited in the middle of a track during brief halts. Male rhinos expel their urine backwards in a powerful jet, which can prove very startling to zoo visitors and has been known to drench the clothes of the unwary. Bulls sometimes belabour bushes with their horns and feet before spraying them with urine. Heaps of dung are definitely not intended to mark out a specific area as the personal territory of an individual animal. Schenkel believes that hook-lipped rhinos, which have such an acute sense of smell, use this expedient to maintain contact and familiarize themselves with an area even when they are out of visual range. The same thing probably accounts for the female rhino's habit of emitting sporadic jets of urine while on the move.

The zoologist Herbert Gebbing investigated the sleeping habits of rhinos at Frankfurt Zoo in 1957. The animals generally lie on their bellies, slightly off-centre, with their forelegs tucked beneath the body and their hind legs extended forwards. The head rests on the ground, facing front. Only in rare cases will an animal lie flat on its flank with all four legs extended sideways.

It may be that the animals sleep particularly deeply in the latter position. Our two rhinos at Frankfurt lay down shortly after their house was closed for the night, usually just after the evening meal. Unlike elephants, they slept for a considerable period – eight or nine hours, on average – and spent almost as much time on their right side as their left. Rhinos normally stay put for two to three hours, sometimes as much as five, and are not disturbed by familiar sounds. Their breathing, which sometimes sounds like snoring, is distinctly audible. Respirations per minute vary between eight and ten. Two or three times a night they stand up to excrete or urinate. The male's urine may travel for twelve or thirteen feet.

Gerda Schütt noted that the rhinos at Hanover slept nine-and-a-half hours a night and stood for an average of three hours, almost all of which time was devoted to eating. If one of them stood up the other soon woke too. If not, the first would nudge the second with its head until it also stood up.

Except at wallowing-places, rhinos are invariably found alone or in small groups numbering five head at most. Pairs usually consist of a cow and a growing calf, sometimes of a bull and a cow, and – rarely – of two bulls. Adult rhinos do not stick stubbornly together for long periods or for good, as Cape buffalo do. Cows never consort with bulls while their calves are still small, only when they are half-grown. Rhinos standing side by side occasionally brush one another with their lips or rub their neighbour with the underside of their chin. One evening in 1958, Game Warden Ellis of the Nairobi National Park saw a group of four rhinos emerge from the trees. Three of these full-grown animals were acting in a peculiar way. They advanced shoulder to shoulder while the fourth walked behind. The cow in the middle showed signs of labour. The animals halted when they noticed that they were being observed, but one of the cows continued to massage the flank of the mother-to-be with her horn and the side of her head. Eventually they all retired into the bush again, and three days later a new-born calf was sighted.

Seemingly hostile encounters between rhinos almost always pass off peacefully. A mother may be standing with her calf when a big bull appears from behind a bush. All heads jerk upwards, the cow snorts, and the bull snorts too. Stiffly, the two giants raise their little tails like alarm-signals. The bull paws the ground with its hind legs and advances a few yards. The two animals are about 80 yards apart now. Once more the bull snorts, and the cow does likewise. Then, almost simultaneously, both animals lower their heads and charge. I raise my camera and focus. Sixty yards, fifty... I can already hear the mighty thud as tons of flesh and bone collide. Thirty yards, twenty... Suddenly, at six paces, they halt and eye one another with heads erect. The trumpet-shaped ears are cocked forwards, the cow swings her head gently to and fro. Then the bull turns aside and walks to the water's edge. Shortly afterwards the cow follows suit, and a little while later all three animals are peacefully rubbing shoulders. Yet another picture of a rhino-battle that came to nothing!

Hook-lipped rhinos clearly acknowledge the superiority of the elephant, though the animals so seldom have cause to quarrel that this is not readily apparent. An elephant and a rhinoceros which were approaching one another at a leisurely pace along a track in Uganda did not notice the fact until they were fifteen yards apart. The elephant spread its ears and headed straight for the rhinoceros, which halted and raised its head. The elephant then charged and the rhino went into reverse, swinging its head from side to side and snorting loudly. Another quick lunge by the elephant put the rhino to flight. It galloped off the way it had come, but shortly afterwards the two animals were grazing not far apart without taking any notice of one another. One day, in an area which now forms part of Arusha National Park, Frau Trappe found a rhinoceros which had been transfixed by elephants' tusks, and round it

How they get on with other animals

the imprint of elephants' feet. Similar cases have been observed on numerous occasions. The game warden Koos Smit told of a fierce battle between a bull rhinoceros and an elephant at Kruger Park in 1960. The elephant evidently wanted to prevent the rhinoceros from drinking but the rhinoceros insisted. During the ensuing struggle the two animals plunged ten feet down the river-bank but continued to fight in the water. Great pools of blood led to the spot where the rhinoceros lay dead with four tusk-wounds in its body apart from other injuries. Elephants have often been observed to cover dead rhinos with branches and twigs.

The relations between rhinos and other big game are very much less well-defined. The game warden of Murchison Falls National Park saw a black rhinoceros chase a group of twelve waterbuck for about a hundred yards. The antelope recovered from their surprise, veered, and attacked the rhinoceros in their turn. The grey giant hastily retreated into some undergrowth and did not reappear. On another occasion, in sheer bravado, a rhinoceros attacked a herd of about 350 Cape buffalo which were grazing in a row approximately a quarter of a mile long. It charged along the line of unsuspecting buffalo, scattering them in all directions, and then continued on its way.

Mutual toleration is not only commoner but may verge on friendship. A. Ritchie tells of two big rhinos which were for a long time to be seen with a large herd of buffalo. The rhinos regularly slept in a clearing surrounded by buffalo and actually lay side by side with them. In Nairobi National Park Guggisberg saw a group of zebras playfully attacking a lone rhinoceros, which eventually moved off.

A black rhinoceros cow was seen wallowing in a river-bed in Natal with two aquatic tortoises tugging at the fissured layer of hide which is so often found on a rhino's flanks. Although this obviously hurt the animal, because it jumped up whenever one of the tortoises tugged too hard, it made no move to attack them. On another occasion, also in Natal, a hook-lipped rhinoceros was seen to lie down on its side in a pool. At least half a dozen aquatic tortoises immediately converged on the animal and began to feed on its ticks, rearing as much as six inches out of the water in order to reach the parasites. They dislodged them by placing their forefeet against the rhino's body, taking a tick in their mouth and pulling hard until the parasite came away. The big animal obviously found it unpleasant when the tortoises were working on the more sensitive parts of its body, because it twitched several times, but the tortoises took absolutely no notice. Although cattle egrets follow rhinos about the whole day long and perch on their backs, their sole concern is to catch the insects which the big animals disturb. Examination of cattle egrets' intestines has shown that they do not relieve rhinos of ticks.

The giant creatures readily turn tail when confronted by unfamiliar animals. Simba, a fox-terrier belonging to that admirable photographer Cherry Kearton, put two big rhinos to flight, barking loudly. The dog pursued them so fiercely that it was later found almost five miles away in a state of utter exhaustion. In the days before jeeps and Land-Rovers were common in Africa, people went after big game on horseback. Kearton often saw horsemen pursued by rhinos. The speed of a charging rhino roughly corresponds to that of a galloping horse, so the chase could be of long duration. Although riders could not build up much of a lead, the pachyderms abandoned the chase after a while because their staying-power was inferior.

Rhinoceros calves are occasionally killed by lions. In 1966, at Manyara National Park, Tanzania, a mother-rhino and her calf were attacked and driven towards the park entrance. The lions caught the calf about fifty yards short of the gate-house. The mother uttered pitiful squeals for help. As it happened, a guide's car was heading out of the park and a visitor's car was just about to enter. Both cars were driven back by the furious female, which could only be prevailed upon to let them pass after much shouting and stone-throwing. The lions left the calf's remains lying there and made off.

Near Lerai Forest in the Ngorongoro Crater, a young adult

Baby rhinos killed by lions

rhinoceros was killed by a pride of lions, which mauled it badly about the throat. No traces of a struggle could be seen on the ground, so it was assumed that the lions had broken the animal's neck. Although the lions remained beside the dead rhino for an entire day they did not attempt to eat it. Next morning they abandoned the carcase and moved on. A lion which tried to kill an eleven-month-old calf belonging to the female rhinoceros Felicia in the Ngorongoro Crater in August 1966 fared quite differently. As soon as Felicia saw her offspring in jeopardy she flew into a frenzy. The lion managed to separate the calf from its mother, whereupon it ran off with the lion in hot pursuit. Felicia brought up the rear, bellowing loudly. The calf circled and came back to Felicia, who immediately attacked the lion. Although the latter caught her by the hind leg and badly injured her thigh, the cow spun round and gored the lion in the ribs. The predator collapsed and lay still, whereupon Felicia thrust her horn into its neck and head and trampled it to death. The whole thing was over in a few minutes. Two other lions, which had been sitting close by throughout the encounter, maintained a respectful distance. Less than forty minutes after its death the lion's carcase had been picked clean by hyenas. A group of German visitors accompanied by Game Warden Shehe were able to watch and film the whole incident.

Rhinos generally take no notice of lions, even when they pass close to them. At Mzima Springs, a spring-fed pool of crystal-clear water in Tsavo National Park, Guggisberg saw a rhinoceros killed by a hippopotamus. The rhino was about to drink when the hippo grabbed it by the right foreleg, pulled it into the water and cut it to ribbons with its enormous tusks. Selous photographed a full-grown rhinoceros cow which had been dragged under and drowned by a crocodile.

When two hook-lipped rhinos do fight, which happens quite seldom, they make an enthralling spectacle. Misled by the example of stags, antelopes or buffalo, people at first assumed that it was always bulls which fought, either for reasons of sexual rivalry or for possession of territory. In fact, fighting generally occurs between a cow and a bull or, less frequently, between two cows, and mock battles are not uncommon. Our two hook-lipped rhinos at Frankfurt Zoo often spar for hours, horn against horn, and similar contests between the calf and its father or mother are still more frequent. Rhinos seldom inflict serious injury on each other, even when fighting in earnest. The commonly observed wounds on shoulders and flank have other causes, as we shall see later.

When a cow is on heat the bull stands facing her and the two animals sniff one another about the muzzle, often emitting gurgling sounds. Then, almost as a matter of course, the cow attacks the bull and butts him violently in the flank. The bull tolerates this, although the impact can be such that it forces a belch out of him. Even if a second bull appears and starts to prance round the cow as well, the two males do not fight but leave the cow to decide which of them to bestow her favours on. This love-play is accompanied by loud snorting, panting, and a sort of grunting. Squealing is also heard. Our bull at the zoo emits a loud and piercing whistle, though I have never heard this in the wild. It probably expresses surprise, and is only repeated at longish intervals. One can even coax rhinos to approach by imitating their snorts and heavy breathing.

The animals mate at all seasons. They also produce young throughout the year. Martin Johnson, who made the first good wild-life films in East Africa and the Congo during the twenties and thirties, once stopped his car to observe the love-play of two rhinos at very close quarters. The animals stamped about, taking short, stiff-legged steps. After half an hour the bull got wind of the car, gave a snort of alarm, and raced off into the bushes with tail erect. 'We naturally thought the cow rhino would do the same, but she didn't. It was almost as if she hadn't seen her suitor disappear. She seemed quite astonished that his courtship had ended so abruptly. Then she became aware of us, and – wonder of wonders! – began her amorous goings-on all over again as if she mistook our motor vehicle for another rhinoceros. That our car should suddenly have become an object of veneration to a cow rhino was a new experience for us – nor was the new love affair limited to a fleeting moment. The vain creature tried to break down our motionless car's dumb reserve for fifteen minutes or more. She coyly withdrew, and nothing happend. She paused, then frisked clumsily about. She flirtatiously plucked a bunch of grass and tossed it into the wind. She strutted daintily towards us and came even nearer than the spot from which she had beaten her unsuccessful retreat. Then, all at once, she scented us. With a furious, outraged snort, the angry creature abandoned all her coquetry. Down went her head, up went her tail, and, so suddenly that we were taken quite by surprise, she charged straight at us. The next moment she collided – fortunately at an oblique angle – with our wing. But we were not the only ones to be surprised. The metallic crash which resulted from her attack and our own shouts were sounds new to her ears. The cow gave another snort of fury and then made off wildly in the direction of the salt-pan.'

We have often observed the actual mating process at Frankfurt Zoo – something which has rarely been seen in the wild. Frank Poppleton has described how he watched a bull rest the soles of his feet on a cow's back and remain in that position for a full 35 minutes. The two heads lay side by side, and the animals very slowly advanced in a circle. When the bull had dismounted the female turned to him and the two animals eyed one another for a couple of minutes. My associate Dr Scherpner watched a mating which lasted between 21 and 22 minutes in Tsavo National Park. John Goddard saw rhinos mating on six occasions in the Ngorongoro Crater during 1964 and 1965, and the process was the same. In one instance cow and bull remained together for four months after mating. Another pair parted shortly afterwards, were seen mating a month later, and then split up again. Mervyn Cowie, formerly director of the Kenya national parks, once saw a bull mount two cows in quick succession, having been attacked by the first cow in the interim.

Now that black rhinos have reproduced themselves in zoos,

Assisting at a rhino's birth

their gestation period is known. It lasts for fifteen months, or 540 days. No instance of a twin birth has yet been recorded. The first black rhinoceros to be born and bred in captivity saw the light at Chicago in 1941, the second was born at Rio de Janeiro Zoo, and the first European rhino at Frankfurt Zoo in 1950. $3\frac{1}{2}$ gallons of amniotic fluid were discharged at the birth of the latter. Our female rhinoceros Katharina die Grosse was so tame that she could be milked even before giving birth. Labour-pains were hard to detect, and the first definite signs occurred 90 minutes before delivery. The cow allowed the veterinary surgeon, Dr Klöppel, to deliver the calf, which weighed about 55 lb. The new arrival's ears moved after a few moments, and two minutes later the mother belatedly attacked the helpers standing around in her stable. She sniffed her offspring but did not lick it. The calf stood up for about two minutes after ten minutes, but it was moving about briskly only an hour after birth, remaining on its legs first for half an hour and soon afterwards for a full hour on end. After four hours it found the udder and sucked. It did not spend any considerable period lying down – an hour, to be precise – until nine-and-a-half hours later. At birth, the front horn consisted of a half-inch protuberance and the future location of the rear horn was marked by a white patch. Other rhinos born in zoos have weighed between 45 lb. and 85 lb. (Hanover.)

To the best of my knowledge, all hook-lipped rhinos so far born in zoos have been reared successfully, two of them by us at Frankfurt. Both at Frankfurt and in Rio, cows have been observed to mate regularly while pregnant and are not separated from the bulls. Our own cow was completely docile with her keeper and all persons familiar to her after only a week. We could enter her stable, ride on her, and play with the calf.

In 1911 the Hungarian explorer Kalman Kittenberger inadvertently shot a female hook-lipped rhinoceros just as it was in the process of giving birth. He slit open the dead beast's belly and brought the calf out alive, but it died a week later. The birth of a black rhinoceros was first observed in the wild by Game Wardens

Rhino-milk

Malinda and Edy at Manyara National Park in 1963. They came upon a female rhinoceros stretched out on the ground. Suspecting it to be dead, they threw a few stones at it. The animal did not rise, so they approached and found that the ground around it was completely saturated. After another few minutes the rhino got to its feet and the calf emerged without seeming to cause its mother any great discomfort. Ten minutes later the calf dropped to the ground. The mother turned and started to remove the caul with her lips. Another ten minutes and the baby rhino was standing up and waggling its ears. Then the game wardens proceeded on their way.

John Hunter, the mass-exterminator of rhinos, wanted to capture a baby calf for sale. Having already dispatched 75 rhinos on the hunting-trip in question, he managed to catch a youngster by luring it with the udder of its dead mother. Rhinos captured at such an early age become very tame, like domestic animals, and even adults soon grow accustomed to the society of human beings. We milked our cow rhinoceros numerous times. Ten months after birth the milk contained 3.2% protein, 36% lactose, and 0.3% fat. The young continue to feed from the mother's two dugs for about two years and remain with her for at least three-and-a-half years.

A cow rhinoceros is not generally covered again until eight to ten months after giving birth. At Amboseli the first calf, the earless Pixie, remained with its mother Gertie for two years nine months, the next calf for three years. The third was born in 1959, after an interval of five years. The animals become sexually mature at about seven. Mothers sometimes lie down to suckle their young like domestic pigs. They never consort with bulls while their calves are very small.

Like reindeer, deer, and many other species of wild game, rhinos become inquisitive about human beings or other suspicious figures whose scent has so far eluded them, and edge gradually closer until they at last run off. Hook-lipped rhinos have another characteristic which has proved very detrimental to them. Snorting with every appearance of fury, they will charge to within a few paces of a shape whose significance they cannot determine, then swerve aside or simply run past. The film-maker Martin Johnson and his wife once jumped off a rocky ledge to escape some charging rhinos, only to find that the gigantic beasts had halted five yards short of the spot where they had been standing. In two other instances, when the Johnsons had no possibility of escape and no chance to reach for a gun, rhinos turned aside just before reaching them. However, few people have the nerve to wait deliberately and see whether the myopic animals are making just another exploratory lunge or charging in real earnest. A hunter will always shoot first.

The grey giants sometimes charge tree-trunks or termites' nests in the same way and then simply pass on. Once, when John Owen, Director of the Tanzania National Parks, was climbing out of the Ngurdoto Crater on foot, accompanied by a noted horsewoman, a rhinoceros unexpectedly appeared on the path ahead of them. Owen hurriedly dived sideways into the undergrowth. His companion leapt for a branch and pulled herself up, but the branch broke and deposited her astride the rhino's back. It would have been hard to say which was the more startled, the rhino or its involuntary rider. In the end, she fell off and the rhino trotted away.

One should not, however, put too much faith in the harmlessness of rhinos. This fact was forcibly impressed on Rudolf Schenkel, who observed rhinos and lions on foot in Kenya's Tsavo National Park and spent many nights in the open with no protection other than a sleeping-bag. Many of his encounters with black rhinos did, in fact, pass off uneventfully. Then a bull spotted his moving silhouette outlined against the evening sky at a range of about fifty yards. Schenkel shouted and ran at the animal to scare it away but was forced to dodge because it continued to bear down on him at full speed. Then he raced to a small tree, half of which had split off and was hanging from the main trunk in a tangle of withered foliage and branches. Having no time to climb into the surviving half of the tree, Schenkel ran round it and vaulted the broken

The professor's ordeal

section of trunk while the rhinoceros had to negotiate the withered top. Soon, however, the bull changed its tactics. While Schenkel hovered on one side of the tree, beside the broken section of trunk, it lurked on the other, ready to launch a sudden attack. Schenkel at last tried to pull himself up into the living part of the tree but was caught and tossed into the air. He landed on the animal's shoulder, slid to the ground, and immediately crawled beneath the dead tree-top. The bull hurled tree-top and snapped trunk aside with a single jerk of its head. Schenkel decided to remain motionless, simply raising his foot to the level of the rhino's snout so as to ward the beast off in case of need. The bull started back at first, then edged closer until its nose came into contact with Schenkel's bare foot – his shoe had fallen off. Now that it no longer saw a moving shape, the bull responded to the scent of human being. It abruptly turned away and trotted off with tail erect.

The behaviour of the hook-lipped rhinoceros thus varies widely according to the type of human being with which it shares its habitat. The Wakamba of Kenya hunt the animals with poisoned arrows or traps. The poison is decocted from the twigs of the Acokanthera schimperi. The fresh poison is very strong but soon loses its potency when exposed to the air, so the poacher keeps his arrow-heads wrapped in a piece of hide until just before use. A rhinoceros wounded by a fresh arrow soon dies, wherever it may have been hit. Wire leg-nooses fastened to a heavy balk of timber are less merciful, and cut deep into flesh and bone. The poor creatures drag them around for weeks on end, which is why the rhinos in Wakamba territory are always reputed to be spiteful and aggressive.

The Masai, who are not hunters and leave the animals alone, regard them as quite peaceable. In 1964 at the Hluhluwe Game Reserve in Natal a game warden was twice tossed into the air by a black rhinoceros and sustained gaping wounds in the thigh and buttock. When the rhino attacked a third time the game warden

Attacks on cars

probably saved his life by grabbing its front horn and hanging on desperately. The rhino swung its head violently to and fro, trying to shake the man off. It eventually succeeded in doing so with a particularly vigorous heave, but the injured game warden flew sideways into the bushes. At that moment the rhinoceros ran off.

Rhinos which attack without warning often turn out to have been wounded previously. Oscar Koenig tells how, while driving from Moshi to Same in Tanzania, he shot a rhino which refused to let him pass and wounded it in the hind quarters. The animal overturned three private cars and two trucks in the nights that followed and had to be shot. Kearton records that a woman hunter was killed by a rhinoceros noted for its friendly disposition after wounding it with a gun of far too small calibre. Next day a local settler drove along the road with his wife. No sooner had the rhino sighted the car than it charged. The settler quickly pulled his wife out of the car and helped her into a tree, but he himself was caught and killed.

I have experienced numerous attacks on cars, all of them provoked by me. In most cases the animals stopped short of the car without touching it. In one instance the vehicle received a dent. On another occasion the son of the game warden at Amboseli drove me, rather too fast, up to Pixie, the earless rhino mentioned earlier, whose auditory canals I wanted to examine at close range. The animal, which was sound asleep, jumped to its feet and promptly attacked us, inflicting a dent in the side of the open car just by my seat.

In 1965, also at Amboseli, a rhinoceros thrust its horn through the open window of a fully-laden saloon car and transfixed the roof, badly denting the body-work and injuring the passengers with the shaft of a spear which was protruding from its neck. The animal was shot by an escorting game warden. In 1958, in the same park, a car-load of visitors surprised a female rhinoceros with her six-weeks-old calf in the bush. The animal, whose front horn was already broken off, charged the car, hurling two women passengers over the driver's head. The latter punched the rhino on the snout.

Handcuffs on a horn

He also shouted and banged the side of the car, whereupon the calf ran off and the mother followed.

In 1966, when a car visiting Serengeti tried to creep slowly past a rhinoceros standing on the road, the animal attacked it, lifting the vehicle slightly and buckling one wing. One of the passengers was thrown head-first through the windscreen but sustained no serious injury. In the Hluhluwe Reserve in Zululand an old cow rhinoceros approached a car, put her head underneath the front wing – clearly without evil intent – and began to rock the vehicle. The escorting African game warden climbed boldly out and hit the animal on the head with his belt, to which were attached a pair of handcuffs. The surprised rhinoceros trotted off and the game warden hurled his belt after it. By a curious coincidence the handcuffs lodged on the animal's horn and remained there until it had gone several hundred yards. The car was undamaged. At Kilometre 555 on the railway line between Moshi and Same, a rhinoceros chased away all the plate-layers and dented their trolleys. It had to be shot.

In former times, the few rhinos displayed in zoos were almost all 'armoured' rhinos from India. These have since become virtually extinct, and only a handful now survive in captivity. The first black African rhinoceros to reach Germany was acquired by Berlin Zoo in 1903, the first to reach Switzerland by Basle in 1935. Today, the genus is most commonly represented in zoos by the black or hook-lipped rhinoceros. In 1966 there were 32 of them in United States zoos alone. A hook-lipped rhinoceros costs more than twice as much as an elephant. In general, the creatures become extremely tame under human supervision, and many adult females allow themselves to be ridden without demur. They are fond of having their closed eyes stroked with the flat of the hand. Presumably for want of something to do, they like to rub their horns against concrete walls and metal bars. This often reduces them to short stumps, which is why no rhinoceros enclosure is complete without a tree-trunk of soft pinewood on which the animals can sharpen their horns. They seldom venture right into pools, unlike elephants, but are fond of taking mud-baths. A ditch 6 feet wide at the top and 4 feet deep at the outer edge will deter them from crossing even if the inner edge of the ditch slopes towards them. Zoos afford our only clue to the life-span of these creatures. At Chicago Zoo the pair kept for breeding purposes were still alive twenty years after their arrival and showed no signs of ageing. It is probable that rhinos are as long-lived as elephants and therefore attain an age of about fifty.

Our ability to immobilize the animals by means of anæsthetic or paralytic darts has made it far easier to capture and translocate them or give them medical treatment. The celebrated cow Gertie had a badly damaged eye removed in this way at Amboseli in 1962, and had virtually recovered after 24 hours.

Numerous conjectures have been made about the wounds, often crescent-shaped, which hook-lipped rhinos in many areas frequently carry to the rear of their shoulders. They were once thought to be battle-scars or wounds that had been enlarged by starlings. While examining four rhinos in Tsavo National Park, Kenya, not long ago, J. G. Schillings discovered that these wounds contained gossamer-fine filarial worms of the sort transmitted by stable flies. In addition, the larvae of bot-flies live in rhinos' stomachs. These are anchored by their heads to the stomach wall, where they feed on blood and lymphatic secretions. As soon as they mature they pass through the anus and hatch out on the ground. The fat-headed flies which emerge from them vary between 2 and 3.5 centimetres in length. They do not ingest nourishment but remain permanently near the rhinoceros and attach their eggs principally to the head and the area round the horns. It is not known how they get from there to the stomach. Two species are involved, Gyrostigma conjugens, which occurs only in the hook-lipped rhinoceros, and G. pavesii, which occurs in the white as well as the black species. In addition, twenty-six different species of ticks have been identified on white and black rhinos, though these are also found on other animals. Furthermore, rhinos play host to a species of parasitic worm which lives in the intestines of ele-

Little rhino-pests

phants too. They also harbour tapeworms which never exceed 7-12 centimetres in length and can measure as little as 1 centimetre. None of these animal parasites is injurious to human beings or domestic animals. What is more, zoo rhinos generally become free from parasites very quickly because zoos lack the intermediary hosts which transmit such pests.

3. What the lion's prey feels

Enter a land where everyone plays the lion, and you cannot play the goat.
AFRICAN PROVERB

Lions have to yield the right of way to an elephant when they meet one in a narrow defile, just as they also give ground before a charging rhinoceros. And yet to most people the lion is still king of the beasts, just as it was in the fables of old. Skilled African hunters who have grown more and more weary of slaughter, as the best of them tend to do, usually give up shooting lions first. Three of my friends have admitted to me that the only time they now pick up a rifle is to shoot a zebra for an old or ailing lion in need of food. It is clear that lions excite our admiration more than other animals. Lions figure in the royal arms of England, Scotland, Norway and Denmark – all countries which have never seen a lion in the wild. They also appear on the arms of Zurich, Luxemburg, Wales and Hessen – once again, places where lions are no longer indigenous. Cave-lion's bones dating from prehistoric times can be found throughout England, Germany, France and Spain. The cave-lion, which was a contemporary of man, probably looked very much like the lion of today. Lions did not become extinct in Greece until circa 200 B.C., and the Bible makes no less than 130 references to their presence in Palestine.

Properly speaking, therefore, lions are not tropical beasts. In Africa their spoor has been found at 11,500 feet on snow-capped Mounts Kenya and Ruwenzori – indeed, at altitudes exceeding 15,000 feet. For once, it is probably wrong to charge the human inhabitants of our more northerly lands with having exterminated

The lion's roar

the cave-lion, because they had not yet succeeded, with the modest weapons at their disposal, in doing the same for the bear, elk, bison and aurochs. The steppe-dwelling lion probably retreated as more and more of our own continent became clothed in forest. Nevertheless, the lion's fame was such that stone effigies of it have stood for thousands of years outside the temples and palaces of China, a country to which the animal has always been a stranger.

What is it that we really find so impressive about the lion? For one thing, the majestic Jovian head and billowing mane; for another, the amber-coloured eyes, which are larger than our own (diameter of the human eyeball, 23 mm., of a lion's, 37.5 mm.). Probably, too, its roar. Lions, tigers, leopards and jaguars have circular pupils and are 'roaring cats', in contrast to the more numerous and less awesome 'purring cats', which have vertical slits instead. The lion's roar has been described as the most magnificent and imposing sound in creation, and can under favourable conditions be heard upwards of five miles away. The lion usually roars in a standing position with its head slightly inclined. The flanks are indrawn and the chest expands mightily, like a bellows. Often, the animal stirs up the dust in front of it by the very act of exhalation. A lion's roar affects me rather like the church bells I used to ring as a boy – it induces a solemn and serious frame of mind. Many people say it gives them an agreeable prickling sensation, but that probably applies only when they are sitting in a car, visiting a zoo, or listening through the open window of a rest-house in the African plains. At Serengeti the lions sometimes gave tongue a couple of yards away from our aluminium hut, or so it seemed to us, and we tumbled out of bed in a hurry. The effect on a man alone in the open on foot tends to be something other than an agreeable prickling sensation or a mood of solemnity. Lions are at their most vocal for about an hour after sunset.

Why do they roar, in fact? No one has yet advanced a wholly satisfactory explanation. Professor Hans Krieg believes that the roar is simply a luxury, like the display plumage of the bird of paradise, like monkeys' games, like the sudden bouts of cavorting indulged in by antelopes, like the choruses sung by howling-monkeys and gibbons or the somersaults of the ratel. On the other hand, lions may intend it as a solemn proclamation to all other lions – and lionesses – that a particular territory belongs to them. What is quite certain is that antelopes, zebras and gazelles do not, contrary to the statements made in many books, race off in panic at the sound.

The lion's family life – with one or two exceptions – makes a very attractive impression. Males do fight among themselves, as witness the traces of blood and large tufts of black or yellow mane-hair that mark the scene of a scuffle afterwards. Fighting occasionally results in death, but probably no more often than it does in the boxing-ring. Lions subscribe to certain rules of behaviour. A pride of lions will feed communally on the same zebra carcase, whereas many of our household dogs refuse to eat from the same dish as others of their kind. Pregnant lionesses choose a spot among the rocks or in the bush to have their cubs, but they readily introduce them to the pride at the age of six weeks, just as the celebrated lioness Elsa brought three cubs to her human friends' camp when they had reached that age. Lions do not eat their own cubs and will only eat those of a strange pride under very exceptional circumstances. For all their snarls and grimaces of annoyance, they allow cubs to play with them and try to tug pieces of meat out of their mouths with their little milk teeth. Hunting is a communal activity, and the sick and weak are kept in food for a long time. No one knows whether old lions are deliberately excluded from the pride or shun other lions' society of their own accord, but old females stay longer with the pride than males. All are eventually torn to pieces by hyenas and wild dogs. No creature dies of disease or old age in the wild.

Mating pairs go into retreat and devote days to nothing but love-play. When in season, they copulate thirty or forty times a day. One pair of lions at Dresden Zoo mated 360 times in a single week. The act itself lasts only between three and six seconds. It seems that male lions range over an area of many square miles, attaching themselves to this or that group of females and young for varying periods of time. This certainly applies to the strongest specimens. Each pride has its own ranking order, but the weakest male invariably ranks higher than any female.

It is always hard to tell what lions will or will not do. Dozens of people have successfuly escaped them by taking refuge in trees for the night, yet I myself have photographed lionesses perched in tall trees with vertical trunks. At Manyara National Park, near Arusha, lions have become accustomed to sitting almost all day in trees, where they can conveniently be photographed by visitors. Perhaps there are fewer tsetse flies at twenty, twenty-five or even thirty feet, or perhaps they can spot their prey better from that altitude, but why should this habit be virtually unique to Manyara?

One big game hunter allegedly measured a lion's leap at forty feet, and Count Telski saw a lion clear a ravine thirty-six feet wide and seventy-two feet deep. On the other hand, zoo lions in open-air paddocks have been living behind ditches well under thirty feet wide for the past half-century, and none has ever jumped across.

Lions have, however, drowned in zoo moats not equipped with some form of egress. Zoo lions are never seen bathing or swimming voluntarily, unlike tigers, which are very fond of water. The Adamsons' lioness, Elsa, sea-bathed enthusiastically with her human companions for hours on end, and at Lake Victoria lions have frequently swum to the island of Ukerere, which is over two hundred yards from the mainland.

Professor Antonius, the Viennese zoo director, asserts that the lion is so lazy that it will not move a step provided its food is deposited in front of its nose. Anyone who has seen these animals in Africa, lying beneath the same bush all day long, will confirm that. On the other hand, lions in areas overstocked with game will roam around for hours on end without any apparent motive.

Although there are white tigers and black leopards, albinos and black specimens in almost every animal species, no one had ever seen a black or white lion until 1962, when a visitor to the Kruger National Park in South Africa not only sighted but filmed a white lioness. Many hunters and explorers have alleged the existence of breeds of brindled lions, but the creatures in question always turned out to be individual specimens which had never lost the brindled markings of their cubhood. Conversely, some litters are born without the brindled coat typical of the lion-cub.

There seems, in general, to be only one species of lion common to India and Africa, although dozens of subspecies were distinguished in the last century on the basis of skulls and skins acquired by museums. The last lions in India, in the Gir Forest north of Bombay, had dwindled to thirteen by 1908 but rose again to two hundred in succeeding decades under strict protection. They do not have yellow manes smaller than those of African lions, despite widespread reports to that effect, but include many variations. The lions of India were not ousted by tigers, either, tigers being forest creatures and lions creatures of the open plain; it is simply that a forest creature is less easily exterminated than one which can be hunted in open grassland. The northernmost and southernmost lions of Africa, the long extinct Berber and Cape lions, were no larger or darker-maned than the rest. Thirty years ago, in a South African club-house, someone found a stuffed lion which had been shot near Capetown in 1836. This animal, which is now in London's Natural History Museum, is scarcely distinguishable from present-day lions. The same applies to the two unique stuffed North African Berber lions which can be seen at Leiden Museum in Holland. At Serengeti and in the Ngorongoro Crater, however, large and small, short-maned, corn-yellow and black-maned lions can be found living side by side.

Almost every rule in the animal world has its exceptions, and this certainly goes for the lion. I said at the outset that lions generally avoid rhinos – in fact I once photographed a rhinoceros in the act of putting a lion to flight. Most rhinos take little notice of lions which enter their field of vision. Nevertheless, one large rhinoceros in the Ngorongoro Crater was seen being teased by a pair of lions. The first lion smacked the rhino on the rump from behind. The big animal swung round in a fury, only to receive another playful smack on the backside from the second lion. A very similar

How the lion kills

form of leonine entertainment has been observed in Kenya. On the other hand, one extremely popular rhinoceros which always loitered near the Ol Tukai tourist camp in the Amboseli Reserve was killed one night by two male lions. It screamed so loudly while defending itself that Game Warden Taberer drove out and chased the lions away with his spotlight. The rhinoceros, whose foreleg was broken, had to be put out of its misery.

Lions kill quickly as a rule. They usually spring at the flank of their prey or pounce on it from behind with the forepart of their body on the creature's back and their hind-legs on the ground. They sometimes put one paw over its face and jerk the head backwards. It is always said that this breaks the animal's neck, though this remains to be proved. More commonly, however, they bite through the throat from below and thus asphyxiate their prey. They also achieve the same result by clamping the nose shut with their jaws. It is not true that lionesses do all the work while the lions lazily look on. 'Attacks' which I staged on a dummy zebra were launched by mixed groups of males.

Lions hunt communally. Two or three members of the pride conceal themselves in the grass at a distance from some zebras or antelopes. The others work their way round behind the prey, then attack from the opposite direction and drive them straight towards their fellow-hunters. This technique is innate, because even the Adamsons' lioness Elsa, which had been reared by human beings and could not have learnt the art of co-operative hunting from them, used to round up giraffes while out walking and drive them straight towards the human fellow-members of her 'pride'. If a tame lion does the same with Cape buffalo, the results for two-legged 'lions' can be unpleasant!

An animal which kills quickly and cleanly runs less risk of being injured itself. Starving lions which attack prey too big and strong for them may have to wage furious battles in which they themselves are worsted. Not long ago, for example, a lion had a couple of ribs broken by a Cape buffalo, and on another occasion a bull giraffe smashed an attacking lion's shoulder-blade. The badly

The feel of a lion's jaws

injured animal stayed with the pride and was allowed to feed even though it could not procure food for itself. Where young lions are concerned, many which have to hunt alone do not know what to do with their prey and slowly torture them to death.

Many people have been fortunate enough to escape death, after being seized and dragged outside by a lion, because their companions or dogs have chased the intruder away. Those who have undergone this experience include Game Warden Wolhuter of the Kruger National Park and the African explorer and missionary David Livingstone, who left a precise account of his sensations at the time. He experienced neither pain nor shock nor fear of death but felt quasi-paralysed and totally indifferent, like a half-anæsthetized patient about to undergo surgery. Much the same must apply to a mouse being carried off by a cat or an antelope in a lion's jaws. Many animals which have been hauled along in an inanimate condition get up and run off when the lion puts them down and its attention is distracted. An acquaintance of mine, the game warden Gordon Poolman, had an experience of this sort. He came upon a lion which had just felled a gnu. The predator ran off, and Poolman climbed out of his field-car intending to cut himself a piece of gnu-meat. Just as he turned to speak to his African driver the latter gave a shout of warning. The 'dead' gnu had jumped up and was about to go for him. He managed to seize it by the horns and jump back into his car.

Lions have been known to venture boldly into a tent, only to dash out again in panic because a crate has fallen over with a crash. On the other hand, a group of lions in Kruger Park used to sit round an iron gong at nightfall while one of their number sounded it from time to time with a gentle paw-stroke. Lions repeatedly – and playfully – used their claws to rip the rubber water-bag hanging on the tent-pole at the entrance to our sleeping quarters. They also carried the microphone outside and pulled the attached cable through the wall of the tent. *Simba*, a book by C. A. W. Guggisberg, another acquaintance of mine, contains not only a large number of personal observations made in East Africa but also everything of

The speed of the lion

note that African explorers and hunters have recorded on the subject of lions. According to many accounts, the lion is capable of a speed of 70 m.p.h., but not even a cheetah, which is undoubtedly faster than a lion, managed to clock more than 44 m.p.h. behind an electric hare on an English greyhound track. The American president Theodore Roosevelt, who travelled in Africa and was a great lover of wild life, insisted that a lion cannot overtake a horse – and an English thoroughbred can attain 40 m.p.h. on the racecourse. A lion almost always abandons the chase if it has failed to catch its prey after fifty or a hundred yards. It only persists if the animal is too sick or too young to travel at top speed. This is why antelopes and zebras do not automatically flee in panic from lions which pass them or are clearly visible. They continue to graze quietly, taking care never to let the predators out of their sight. Cape buffalo and elephants have been known to drive lions away from fresh kills and stand over them for hours to prevent the big cats from returning.

It has long been recognized that there are no 'destructive' creatures in the natural state. All the same, it is interesting to know how often a free-ranging lion has to kill to support life. Lions in zoos receive a daily ration of 13-18 lb. of meat, usually horsemeat, beef or whale, all of which suit them well. Tigers, leopards and other feline predators, especially cheetahs, fare better if their diet is interspersed with other kinds of meat, but free-ranging lions feed a great deal on zebras, i.e. wild horses. A lion in the wild can devour up to 40 lb. of meat at a sitting – over 60 lb. according to another estimate. This means little, however, because it does not have to eat every day. On the other hand, the lion often kills more than it can eat. When it cannot cram any more down its throat, it goes off to digest a meal in the shade of the nearest bush, a large proportion of its prey being taken by vultures, jackals and hyenas. The lion, in turn, frequently drives hyenas off their prey. Wells has estimated that a lion kills nineteen head of game averaging 250 lb. each year, or approximately 190 head in ten years of active existence. Wright spent many months observing prides of lions in the open grass-

'Pointsman up telegraph-pole'

lands of East Africa. They travelled between 2,000 yards and 6 miles a night, and three-quarters of their prey – 60% of which were male animals – consisted of gnus, zebras and Thomson's gazelles. On average, a lion killed 13% of its own body-weight a day in the form of wild game. The number of big yellow cats present in a particular area did not depend solely on how many prey-animals were grazing there, but, even more, on the length of the grass. That is why the herds of gnus and zebras at Serengeti move to areas where the grass is quite short when they produce young, as do the Thomson's gazelles with their black-and-white-striped flanks. Short grass cannot conceal a lurking lion.

No need here to repeat the old stories of the man-eaters which delayed construction of the Uganda Railway for weeks on end because they snatched one plate-layer after another and eventually took the engineer out of his wagon. The railway company has preserved some telegrams from local Indian station-masters. The following are dated 1905: 'Lion on platform. Please instruct guard and driver to enter carefully and without signal' and 'Pointsman up telegraph-pole by water-tank. Train to halt, pick him up and proceed.' Believe it or not, the last telegram of this nature dates from 1955.

About a thousand lions live in Serengeti. According to recent research by George Schaller, roughly half the young fail to survive. Starving mothers have even been known to drive their offspring away from meat. Fights between lions, usually of different prides, are very rare but do in isolated instances result in death. A pride's territory, which the males stake out with jets of urine, may cover 100 square miles. Localized prides do not follow the huge herds of antelopes and zebras into the open plain. This is only done by small groups of 'nomadic' lions which have been unable to acquire territories of their own.

Shooting lions really called for courage and skill in the old days. The great hunter and explorer Frederick Selous, who roamed Africa from 1870 until almost the end of the century, started out with a veritable blunderbuss weighing 10 lb. and firing a ten-bore bullet which had to be blasted out of the barrel by more than three-quarters of an ounce of powder. Having wrapped the bullet in a piece of waxed canvas, he cut the folds off with a pair of scissors, rolled it between his palms, rammed it down the muzzle with a ramrod, and then placed a percussion cap in position – all on horseback and sometimes at full gallop. Courageous men of this type ran an even greater risk of dying far away in the bush because quite minor injuries could result in blood-poisoning. Their fame and prestige as explorers and lion-hunters continued, in subsequent decades, to feed the ambitions of many who hunted the big yellow cats by car, equipped with modern weapons, ice-boxes, luxurious tents and crates of canned food.

Like all the great 19th-century hunters who were genuine lovers of nature, Selous would very probably go hunting with a camera today – or, at least, would have done so during the latter half of his life. At about the turn of the century, Karl Georg Schillings roamed the foot of Kilimanjaro with flashlight and rifle, laboriously using his heavy, unwieldy and imperfect camera to take the first photographs of lions devouring a dead zebra by night. Today, at the very same places in what is now the Amboseli Reserve, tourists stop their cars a few yards from tame wild beasts and take magnificent colour-pictures in radiant sunshine. Even so, really good photographic trophies of lions are still rare. You can photograph lions devouring their prey in any position – sitting, standing or lying – but there are very few pictures which show them stalking and making a kill. The process is too infrequent and too swift.

One remarkably well-defined photograph, which showed lions in the act of seizing a wildly rearing zebra, was continually reprinted throughout the world for a period of decades. It transpired, rather belatedly, that an itinerant photographer had removed a whole group of stuffed animals from a small museum in Rhodesia and arranged them in front of a sunlit backcloth of bushes...

So there are still lion trophies to be won in Africa.

4. Do man-eating snakes exist?

Women are like birds: they know all and say little; men know nothing and say much.
AFRICAN PROVERB

A giant snake, eighty or ninety feet long, lurks in the branches of a tree, half-stuns the unwitting man beneath with its bony head, coils round him in the twinkling of an eye, and – always provided that courageous rescuers do not hack it to pieces in time – pulverizes every bone in his body. Hair-raising incidents of this kind figure in many books of adventure and even in so-called eye-witness accounts of expeditions to the tropics.

Do giant snakes really attack human beings, and can they devour us? Few other creatures have been the subject of as many myths as pythons, anacondas and boas. Equally, few other creatures make it as hard, even for the expert, to distinguish between fact and fiction in individual cases.

The first snag is their length. Even serious travellers report that the Amazonian jungle is supposed to contain anacondas 100 feet long or more, but they omit to say if they have seen and measured such creatures themselves or have merely heard tell of them. The anaconda is a South American boa, and probably the longest and most powerful of all giant snakes. Another very famous South American boa, the boa constrictor, attains a length of 'only' fifteen or twenty feet. Another point is that giant snakes are far from easy to measure. The easiest way, of course, is to measure them when fully extended, but this is an unnatural position for big snakes to adopt – in fact many of them are incapable of doing so and have at least to bend their tails. So the powerful creatures will

Not poisonous

not voluntarily stretch out to be measured. If they are dead, rigor mortis usually intervenes before any measurements can be taken. Measurement of skins alone, which dealers handle on a footage basis, always leads to exaggeration. A fresh snake-skin can be stretched by at least 20% – some say 50% – and snake-hunters take advantage of this. Not only skins but live snakes are sold by length. Animal dealers ask zoos somewhere in the region of five shillings an inch for small and medium pythons. The New York Zoological Society has for many years been offering $5,000 for an anaconda thirty feet long or more, but no one has yet produced one. It is probable that huge creatures as big as this or even bigger do exist or did until recently. Big snakes are, however, extremely heavy – an Asiatic reticulated python with a known length of 28 feet weighed more than 250 lb. – and cannot easily be overpowered in the interior and transported to an airfield or port without the assistance of a sizeable team of helpers. The record length for a rock python, a species widely found in Africa, is 32 feet. The Indian python (Python molurus) grows to a length of 21 feet or more, while the East Asian reticulated python (Python reticulatus) attains $27\frac{1}{2}$ or 32 feet, depending on which source of information one chooses to believe. The amethyst python does not attain quite this size. We have now listed all six of the world's biggest snakes, the four egg-laying pythons native to the Old World and the two viviparous boas of the New. There is a whole series of other boas and pythons, but these are smaller, as are the 2,500 other species of snakes which inhabit our world.

Big snakes are not poisonous. Although the poisonous African mamba can grow to 13 feet and the king cobra of India attains an even greater length, both are thin and slender in comparison with the heavy-weights of the snake world. A big snake takes time to acquire such proportions. At Pittsburg Zoo, a 26-foot reticulated python grew 10 inches every year. The older the creatures become, however, the slower their rate of growth.

It is quite impossible to tell by outward appearances whether a big snake is male or female. A pair of rock pythons which were one

Highest recorded ages

year old when they arrived in New York grew uniformly for the first six or seven years. Then the female began to lag considerably, the reason being that she had become sexually mature and was laying eggs every year. This meant that she fasted for six months out of every twelve, both while the eggs were developing inside her body and during the incubation period.

We naturally do not know what age large snakes attain in the wild because no one has yet tagged them in the way that people have been ringing wild birds for decades past. Consequently, we have to base our estimates on the age they attain in zoos. The current record is held by an anaconda which lived in Washington Zoo for 28 years, from 1899 until 1927. A boa constrictor survived at Bristol Zoo for 23 years 3 months, a rock python at the same zoo for 18 years, an Indian python at San Diego Zoo in California for 22 years 9 months, and two East Asian reticulated pythons in London and Paris for 21 years each.

Giant snakes are the only large creatures on earth to be mute, like the rest of their breed. The most they ever manage is a non-vocal hiss. They are also deaf, and cannot pick up the air-borne vibrations which we and other creatures perceive with our ears. They are, on the other hand, very sensitive to vibrations on the ground or other surface on which they may be lying. The deaf-mute giants do not see well, either. Their eyes are comparatively immobile, they have no eyelids, and the transparent cornea protecting the eyeball peels off like an hour-glass at every moult. A snake's eye has no iris-muscles, so the pupil cannot be narrowed or widened in bright or dim light and the eye is virtually unadaptable. Similarly, snakes cannot flex the lens of the eye as we do in order to focus sharply and at will on objects that are distant or close at hand. To do this in some degree, the snake has to advance or retract its entire head. This may be all very expedient for swimming and, more especially, for seeing under water, but the animal kingdom can certainly boast far better eyes than those of the snake. Because pythons, like other snakes, do not close their eyes while sleeping, it is hard to tell whether they are asleep or awake. Some authorities

Deaf and dumb

assert that sleeping snakes direct their eyes downwards so that the pupils coincide with the lower 'lid', but others dispute this.

It was the rigidity of the snake's eye which probably gave rise to the oft-repeated myth that snakes paralyse or hypnotize their prey with a fixed stare. Frogs, lizards or small rodents often remain quite still in the presence of big snakes, it is true, but this is partly because they do not realize their predicament and partly because immobility helps them to escape detection. A snake is more likely to strike at a frog that hops away than one which sits still.

How does the deaf-mute and myopic snake find any prey at all? The answer is that it has senses which are lacking in us – a peculiarly well-developed sensitivity to warmth, for instance. Big snakes can detect the proximity of a human hand from a foot away. This enables them to find hidden prey, particularly of the warm-blooded variety, while gliding about in search of food. So that their own breath should not interfere with this activity, pythons' nostrils direct it upwards and to the rear. The snake's sense of smell is even more acute. Curiously enough, the requisite organ is situated in the roof of the mouth, whither small samples of air are conveyed

Page 81:
Ostriches were long ago exterminated in North Africa and Arabia for the sake of their plumage. They would probably have become totally extinct if people had not learnt how to breed them on special farms in South Africa.

Page 82:
The African spur-winged goose (Pectropterus gambiensis) is really a large long-legged duck. The sharp spur on the wing-joint, clearly visible in our photograph, can be genuinely dangerous. At Dresden Zoo in 1936 a female spur-winged goose struck a woman employee in the eye with its wing-spur and inflicted a cut on her eyeball, only 4 mm. long but fairly deep. The wound did not prove serious, but only because it escaped infection. Spur-winged geese live gregariously beside large expanses of water. They graze on the grass bordering lakes or rivers and sometimes congregate in large numbers.

Page 83:
These three steenboks, a horned male and two females, have retired into the shade of a termites' nest at Etosha National Park, South-West Africa. Steenboks are highly exceptional among antelopes in that they habitually use their hind legs to strew dung with earth like dogs or cats. It seems that they even dig holes for their excrement. They race off when disturbed, sometimes bounding high into the air, but soon stop, lower their heads, and go to ground in hollows in the tall grass or scrub. This conceals them from view in an abrupt and seemingly unaccountable manner.

The one that got away

for examination by the tongue. Snakes are thus independent of daylight and can trail and overpower their prey with equal facility by night or day.

One afternoon, not far from Serengeti, my son Michael and I came across a rock python ten or twelve feet long, and decided to take it with us. Big snakes are not, in fact, very hard to capture provided they do not happen to be lodged in trees or bushes. They would probably cover one mile in the course of an hour – if they ever bothered to travel uninterruptedly for that length of time. Unlike so many smaller snakes, which propel themselves forwards by means of lateral movements, large snakes locomote with their belly-scales. Muscles proceeding from the ribs, which remain rigid, propel the scales back and forth like small grabs or anchors.

We began by cautiously steering the python with forked sticks which we hurriedly cut from the surrounding bushes, and eventually caught it by the tail without being attacked. Having guided it into a sack, we tied the mouth and deposited it under the bed in our hut for the night. Next morning the sack was empty. The big snake had escaped, but we easily discovered its whereabouts by following its track in the dust. It had left a fairly straight furrow as wide and deep as a tyre-track.

No snake, not even a poisonous one, can keep pace with or overtake a running man. On the other hand, big snakes are excellent swimmers – far better than other terrestrial creatures – and the anaconda is almost more at home in water than on land. They are not averse to swimming in the sea, either. A boa constrictor once swam 200 miles from the South American mainland to the island of St Vincent and arrived in perfectly good condition. The volcanic eruption on Krakatoa in 1883 destroyed all life on the island. In the ensuing years and decades, biologists noted how various creepers, plants and living creatures gradually re-established them-

Page 84:
Leopards make a habit of hauling their kill into trees to avoid having to share it with lions, vultures, hyenas and jackals. At Serengeti in 1968, a leopard was itself hauled many feet into the air by a glider's tow-rope.

selves there. Rock pythons were the first reptiles to regain possession of the island in 1908.

Big snakes have not transformed themselves into living ropes as efficiently as most other serpents. Boas and pythons still have two pulmonary lobes in their bodies, like us. In most other snakes the left lobe has completely disappeared; by way of compensation, the right lobe has become extended and considerably elongated. Big snakes also have vestiges of a pelvis and femora concealed beneath their skin, but all that remains of their hind legs is a pair of puny little claws on either side of the cloaca.

How do such slow-moving giants catch their prey? Their secret, I believe, is patience. In the first place, it is untrue that they can knock out a human being or an animal with their heads. The head of a big snake is not particularly hard – certainly softer than our own – so the snake itself would not take kindly to using it as a boxing glove. Apart from that, the creatures do not strike particularly fast. A big snake weighing 275 lb. packs roughly the same punch as a 45 lb. dog. This is enough, of course, to fell a somewhat unathletic European, but a reasonably agile man should be able to defend himself against a 13-foot boa provided he stays on his feet. It ought to be possible for him to push the coils downwards with both arms, using the force of gravity.

Much more important to the snake than 'striking' is an ability to seize its prey and hold on. For this purpose, it opens its mouth extremely wide. A reticulated python has approximately one hundred teeth arranged in six tiers. A man caught by a single finger cannot simply pull it out. He must try to force open the jaws and thrust his hand still farther into the snake's mouth before disengaging it.

A snake does not envelop prey in its coils until it has obtained a firm purchase with its jaws. To handle a giant snake successfully, therefore, one must always grip it just behind the head to prevent it from biting. Take a closer look at film sequences or photographs which show a man 'wrestling' with a giant snake and being choked or throttled by it, and you will almost invariably find that the

They kill by asphyxiation

human victim is holding the snake by the throat. The truth is that he has wrapped the creature round his own body and is simulating a fierce struggle.

Even when a big snake has gripped its prey in its teeth and rendered it immobile by encircling it several times, there is no question of its 'pulverizing every bone' in the animal's body. Not even big snakes weighing several hundredweight possess anything like the immense strength attributed to them. The larger and heavier a creature, the more its strength decreases in relation to its bodyweight. In terms of weight, a flea is ten thousand times stronger than an elephant. Smaller snakes can therefore constrict prey suitable to them far more powerfully than their giant cousins.

Giant snakes kill by inducing asphyxia, not breaking bones. They constrict the chest of their prey to such an extent that it cannot breathe, and the pressure possibly paralyses the heart as well. Once round its victim's body, the snake's coils function more like a rubber cable or cloth than a rope. It is almost impossible to break strong bones in this way, and the many accounts of men's skulls being cracked by snakes are sheer nonsense. A human skull is a hard nut to crack, certainly when the implement used on it is a soft one. My associate Dr Gustav Lederer, who was in charge of our exotarium for forty years, examined three pigs, three rabbits and three rats which had been killed but not yet devoured by big snakes of suitable capacity. Not a single broken bone was found in any of the carcases (though broken bones have been found by other researchers in prey that has already been ingested).

Needless to say, no undue reliance should be placed on these facts. Large numbers of big snakes are kept in zoos. They are not normally aggressive if left in peace, and very quickly become docile. Even free-ranging pythons usually confine themselves to biting in self-defence when attacked and very seldom coil round an adversary. This technique is used only on prey which it is intended to devour. All the same, newly arrived snakes do have to be moved around in zoos, and force has sometimes to be used in order to administer medical treatment. In such instances, one man is reckoned to be strong enough to handle a yard of snake provided he hangs on tight and does not let go.

Despite frequent inquiries, I have never yet heard of a case in which a human being was killed by a big snake in a zoo. Once, decades ago, a senior keeper at Ruhe's, the animal dealers, was enveloped by a 25- or 26-foot reticulated python and allegedly sustained several broken ribs. Similarly, a former snake-dancer told our keepers at Frankfurt Zoo that one of her charges had once broken two of her ribs – not that superhuman strength is needed to do that, least of all to a girl. One of my sons cracked a girlfriend's rib by embracing her too enthusiastically.

Although big snakes are comparatively easy to tame, the ones which dancers wind round their arms and shoulders in cabaret acts and circuses do not need to be tamed at all. Being creatures of variable body temperature, they have only to be cooled before each performance. They will then tolerate almost any kind of treatment and do not become really active until they have warmed up again. Naturally enough, snakes of this kind do not thrive on constant travel, especially in winter, nor on being accommodated in theatre dressing-rooms or inadequately heated hotel rooms and guards' vans. The consequence is that many snake-dancers run through a sizeable number of pythons during their career.

Big snakes do not cling to trees with their tails so as to catch their prey and prevent them from getting away. They do not salivate over dead animals before devouring them, either. This assertion, which recurs again and again, probably stems from the fact that they quite often regurgitate prey which they have already swallowed. This can happen either because it proves too big for them or was not in the right position for ease of swallowing, or because horns or an antler prevent it from going down, or quite simply because the snake has been startled or disturbed. It is only natural that regurgitated prey should be moist with saliva. Even very large and heavy snakes can glide with relative ease through openings such as small windows or cracks in walls in quest of poultry, pigs or goats. With undigested prey forming a huge knot in their belly, they may not be able to make good their escape through the same aperture, so the stable or hen-house becomes a prison. They do not seem 'intellectually capable' of regurgitating the contents of their stomach in order to creep out through the same hole. Several cases of self-incarceration are on record.

People are naturally more inclined to take notice when they come across a snake with a huge swelling in its body – in other words, one which has just devoured a fairly large animal. They are also prone to photograph and write about such an encounter because snakes in this condition are unwieldy and helpless. When an anaconda has a few fish or an adolescent python a few frogs, rodents or birds in its stomach, no one wastes a word on the subject. This has given rise to the idea that giant snakes live on far larger animals of prey than is actually the case. When all is said and done, they are exceptionally modest eaters and can fast for a surprising length of time. The biggest prey tackled by giant snakes of great strength are, say, antelopes the size of a medium roe deer, or pigs – and not the huge domestic pigs of Europe, either, but the wild and domestic pigs of warmer climes. Thus, when people talk of large snakes attacking domestic cattle or large antelopes such as kudu, topi, waterbuck and eland, they are always referring to the young of these species.

The Toro Game Reserve in the Semliki Valley, Uganda, supports about 12,000 Uganda kobs. These antelopes appear to form part of the local rock pythons' diet. Five instances of their having been killed by these snakes were encountered by chance in the course of a single year. The animals concerned were all immature females. Closer examination showed that no bones had been broken, and that they had evidently died of asphyxia. Vultures tried to appropriate part of the prey, and in one such case the python hissed loudly and struck at the birds to hold them at bay. The snake never managed to catch a vulture properly, whereas the birds did succeed in tearing strips of flesh off the antelope. In one instance a 15-foot python weighing 120 lb. had already begun to swallow a female kob weighing approximately 65 lb. The head and neck had already vanished down the throat of the snake, whose body was still coiled several times round the antelope. The python did not move when first approached by P. Martin and the game warden P. Hay, but when one of them tried to remove some tufts of grass from round its head in order to get a better photograph it hissed and swiftly regurgitated its prey. The big snake did not, however, make any attempt to drive the two men away and left its coils wrapped round the antelope. In the Kariba Dam area of Zambia a rock python caught a full-grown monitor by sinking its teeth into the soft portions of the neck. It then wound its body three times round the big lizard, which repeatedly lashed the water with its tail. The Nile monitor was 5 feet long and died soon after being released, whereas the python was almost 8 feet long and uninjured. In another case a $6\frac{1}{2}$-foot python was observed in a tree, coiled round a 3-foot rock monitor (H. Roth).

A snake can swallow another snake of equal length by doubling it up internally. Sandy Terreblanche once watched a small python throttle a large black mamba. The mamba put up a fierce struggle but was dead within two hours.

Many species of snakes have actually adapted themselves to preying upon other species, but very few of them are 'cannibals' proper – in other words, they do not kill members of their own particular species. The stomach of one 18-foot python yielded the remains of a leopard which had only managed to inflict minor injuries on its adversary, though we are not told whether the leopard was full-grown. Our 26-foot reticulated python at Frankfurt Zoo never managed to devour prey in excess of 120 lb. bodyweight, but a 24-foot Indian python devoured a domestic pig weighing 120 lb. and, on another occasion, an Indian long-eared goat weighing 104 lb.

It was the swallowing rather than the killing which presented the latter snake with extreme difficulty. Two days after devouring the pig it was still in such a distended condition that severe injury seemed inevitable. All the other large reticulated pythons kept at Frankfurt over the past few decades have generally refused prey-animals of this size. Although they have been known to attack and kill animals weighing 65 lb. or even more, they often prove incapable of devouring them. Dr Lederer noted that one extremely voracious snake measuring 23½ feet failed, after an hour's struggle, to swallow a goat weighing 75 lb. Another python approximately 25 feet long toiled vainly to devour a pig weighing 95 lb. No expert has ever claimed that a giant snake can devour prey in excess of 130 lb.

If seizure and kill are quickly disposed of, the predator takes its time over the next stage. It releases the carcase, and finally draws itself over the kill like a stocking, beginning at one end, usually the head. Rests are interposed, often for as long as fifteen minutes. A snake can unhinge its upper and lower jaws so that they are connected only by bands. This enables it to open its mouth extremely wide. The upper and lower jaws, with their rows of teeth hooked into the prey, are then advanced in turn. The larynx is also forced forwards so that the creature can continue to breathe. A snake's elasticity extends only as far as its stomach. The other intestines are narrow, so everything must be broken down by the time it reaches them.

Even though pythons and boas gulp down such enormous quantities of food at one sitting, it would be unfair to describe them as voracious. A single meal can provide them with four hundred times as much energy as they expend in a single day, and their ability to fast – of necessity or on impulse – is correspondingly great. One of our reticulated pythons at Frankfurt took no food for 570 days, then ate for a period, then fasted for another 415 days. A Gabon viper (a smallish poisonous snake from Africa) refused food for 679 days, or nearly two years, and an Indian python did not eat for 149 days and lost 10 per cent of its body-weight.

All these facts enable one to draw fairly sweeping conclusions about whether or not human beings are ever killed and eaten by pythons. In zoos, a degree of trust and affection grows up between big snakes and their keepers after a certain length of time. The giant reptiles become accustomed to the sight of their keeper cleaning and moving about in their quarters, and no longer attack. Many snakes, however, remain aggressive throughout their lives. Any sudden movement on the part of their keeper can spark off an attack, even an abrupt movement of the eyes. Once a snake has seized part of a man's body, even if lightly clothed, it envelops him in its coils, as we have been able to observe on more than half a dozen occasions. If it bites into clothing alone, e.g. an overcoat, it will not coil. A trained keeper can deal with a healthy python measuring 14 or 15 feet in length, but healthy snakes more than 20 feet long can be very dangerous to man.

For all that, authenticated cases in which wild snakes have killed and even devoured human beings are almost unheard of. It should be remembered that big snakes often live very close to human dwellings, particularly in remote parts of the Far East. They enjoy great popularity as devourers of rats and cannot, until full-grown, be a menace to domestic animals and human beings. Not long ago a farmer reported in an African trade journal that a four-year-old African child had been going down to the river every day with a bowl of milk and gruel to play with 'Nana'. One day the child's father followed and saw it playing with a large python, which he immediately killed. The whole story strikes me as thoroughly suspect because pythons never touch milk and gruel, but it is an almost ineradicable superstition that snakes drink milk and occasionally suck cows' udders.

In the Napo River in Ecuador a large anaconda once seized a man who was bathing, dragged him down and drowned him, but did not swallow him. A thirteen-year-old boy was also drowned in a tributary of the Napo and then devoured, but regurgitated. His father killed the snake thirty-six hours later. Another authentic case concerns a fourteen-year-old Malay boy on the island of Salebabu, who was devoured by a reticulated python. A veterinary surgeon from the Dutch East Indies who joined Frankfurt Zoo in the early twenties reported a similar incident and even had some photographs of it.

Just how exceptionally rare such cases are, in comparison with the many dramatic accounts that appear in books, becomes clear when we reflect how many large snakes there are in the world – or were until very recently. One indication of this is the number of skins sold. Snakes are not slippery and slimy, as many herpetophobes believe, but pleasantly cool and dry. A snake can travel through mud and water without getting wet or dirty. It glides over rocks on its belly without injuring itself. Now that tanners have learnt how to process unusual skins properly, snake-skins have become much sought after, especially for fashion accessories, though nobody has yet succeeded in preserving a snake's colourful markings in all their original splendour. Trade statistics published by most countries simply refer to 'reptile skins', which include those of alligators, crocodiles, large iguanas and similar creatures, in addition to snakes. Nevertheless, the United States imported 8 million reptile skins in 1951 and Great Britain 12 million. Roughly half would have been snake-skins from the larger, almost invariably harmless snakes, not from poisonous varieties. Roughly 12 million snake-skins are bought and sold throughout the world every year – the equivalent of a snakeskin belt long enough to encompass the equatorial girth of the entire globe. In view of the vast abundance of snakes in warmer regions, the handful of fatalities caused by large snakes may be dismissed as accidents. Certainly, human beings cannot be numbered among their regular prey.

The converse does not hold good. Snakes are undoubtedly eaten by numerous human beings. Madame de Sévigné wrote at the end of the 17th century that her blood had been much refreshed, purified and rejuvenated by the consumption of viper's meat. Snakes – principally cobras – are most commonly eaten in China. Pythons only occur in South China but are imported from other regions. Even in the United States, rattlesnake meat is canned and sold as a delicacy. While hunting in Borneo, Henry Raven reported that the members of his Dyak escort succeeded in killing a 26-foot python just as it was gliding into a river. There were two small pigs inside the snake, so the tribesmen's feast included a bonus of pork. Large snakes, rock pythons in particular, are also eaten in Africa.

Even vultures have been known to overpower big snakes. On a bare, sun-baked and treeless plain near Ngoma, the game warden J. Shenton came upon a python which was being attacked by a circle of eight vultures. They hopped over to the snake, tore at it with their beaks, and promptly retreated before the infuriated creature could strike at them. The python was badly injured. It had great rents in its body through which ribs and intestines could be seen, and one of its eyes had been pecked out. The game warden killed the unfortunate creature, which was otherwise in good condition and displayed no old wounds.

A large python which emerged from under the wing of a car was responsible for a fatal accident near Machadodorp in the Johannesburg district of South Africa. The snake slithered towards the wife of the driver, who let go the wheel in an attempt to save her from being bitten. The car left the road and ran over a local African, killing him. The snake took refuge in the chassis of the car while everyone was busy with the accident and the police were taking statements. It was impossible to get a shot at the creature, so the car was driven to Halfway House in the Transvaal Snake Park, where the proprietor and his staff took three hours to remove the offending python. It turned out to be six feet long and was unhurt.

At Serengeti a leopard killed a fair-sized python measuring 10 feet in length. It perched in a tree with the snake until disturbed by photographers, when it climbed down, still with the carcase dangling from its jaws, and concealed itself in the grass. Later, it climbed back again.

Boas are viviparous. In other words, mothers-to-be retain eggs inside their bodies and incubate them internally, so to speak. The young then emerge fully-developed, as in the case of many species of fish and reptiles. A $17\frac{1}{2}$-foot female anaconda in a zoo produced 34 young each measuring $27\frac{1}{2}$ inches. Pythons, on the other hand, lay eggs in batches varying between 20 and 70, the average at Frankfurt being 46. Freshly laid python-eggs are pliable and shiny, also soft and sticky. The gloss disappears after a few minutes and the eggs adhere together, which reduces their total surface area very considerably and cuts down the rate of evaporation. After only a few hours, the skin of the eggs becomes parchment-like. Pythons' eggs need warmth and moisture but die if placed in water even for a short time.

Pythons hatch eggs by coiling their bodies round a clutch and bedding their heads down on top of it. Paris Zoo discovered as long ago as 1841 that these cold-blooded snakes actually warm their eggs in this way. Experiments conducted with accurate measuring instruments at Washington Zoo in recent years indicate that the brooding female rock python boosts its body temperature by 3-4°C., the male being the same number of degrees cooler. Measurement of the temperature between two adjacent coils of the brooding female's body frequently indicates that they are in excess of 7°C. warmer than the surrounding air. The mother-snake remains coiled round her brood for about eighty days without eating.

Young pythons at Frankfurt shed their skin five to nine times a year, whereas full-grown snakes do so only three to seven times annually, starting at the head. It is possible, with care, to remove this transparent skin intact. If we human beings shed our skin all at once like snakes instead of very gradually, flake by flake, we should no doubt surround our discarded hide with special rites and superstitions. At the very least, we should have to endure half a dozen television commercials every evening for creams and ointments designed to accelerate the skin-shedding process and beautify the new skin beneath.

Even big snakes are sometimes grateful for help in shedding skin. Not long ago in the Transvaal, J. J. Marais noticed a number of grazing cows busily licking something on the ground. Going closer, he saw that it was a large python in the process of moulting. The big snake lay at full length on the ground and was having its skin 'licked off'. As soon as it registered the man's presence it retired into a hole in the ground.

How they mate

When large snakes are five or six years old they go in search of a mate. The males follow the females' trail, evidently informed of their sex by the scent of special glands in the cloacal region. When two prospective mates meet, they thrust their heads together and lick one another. Then the male snake mounts the female and their cloacae unite. At Frankfurt Zoo, copulation has been known to last for as long as two-and-a-half hours.

There is no evidence that earlier periods in terrestrial history produced snakes which were longer and stronger than the large snakes of today. In contrast to saurians and other reptiles whose great days on earth are long past, it seems probable that snakes did not attain their present variety and number until comparatively recently. It is also probable that man first encountered large snakes in Africa, which recent research points to as the cradle of the human race. We do not appear to suffer from a natural aversion to them – that is to say, the fear of snakes is not innate in us. Neither human babies nor young monkeys show any fear or horror of snakes before the age of two, but play happily with them. Interest in these peculiar earth-bound forms of life increases until the age of four. Fear only sets in later, apparently transmitted by adults.

We have made snakes not only into devils, as in the Biblical account of Creation, but also into gods – in fact this is almost the rule where large snakes are concerned. Dahomey used to have priestesses who worshipped the python as a god and carried it in procession. Anyone who killed a python was locked up in a hut, which was then set on fire. If he managed to escape unaided he received no further punishment. The protection of pythons was specially written into treaties concluded between Nigerian chiefs and the British. A European who had killed a python in his own house was strung up by the thumbs, spat on and stripped naked by incensed Africans, but the colonial government deemed it wiser not to undertake any special punitive measures. Reports from districts where rock pythons are considered to be sacred and tabu to the hunter do, nonetheless, make mention of some instances in which small children have been killed and devoured by them – also, in one case, a sick and infirm woman living on an island in Lake Victoria.

There have been python-worshippers in Dahomey ever since one 19th-century king chose the big snake as his royal symbol. Even in the southern, Christianized, part of Dahomey, the inhabitants demand fines for pythons killed on the roads. Ouidah, some 20 miles east of Cotonou, is a Mecca for python-worshippers from all over Africa. The snakes are particularly common here. An American who captured and exported 1,265 regal and rock pythons in a single year, 1967, ran into considerable difficulties as a result. Local Africans threatened to set fire to his house at Cotonou, where he kept the snakes, so he was compelled to build a new one outside town. The neighbours demonstrated in front of it, hurled stones through the windows and tried to overturn his wife's car. They also plastered his house with posters and threatened his native assistants.

Numerous myths and legends are associated with this form of cult. According to them, giant snakes kill bulls for preference and spare cows, whose udders they squeeze dry of milk. In Nepal, they are even said to do the same to women who happen to be suckling their young. A big snake which boarded a ship squeezed a water-barrel so hard that the iron hoops dropped off and fell to the deck. It is also said that large snakes sometimes swallow their own young in order to save them from enemies. A mission journal even recommended readers in danger of attack by a python to lie rigid on the ground and wait calmly until the snake had sniffed them and started to engulf their legs. Once it had reached knee-level, they were advised to pull out a knife and slit the sides of its mouth. The tribes living on Mount Meru in Tanzania believe that a python spits out a precious stone just before it expires. Failure to find this stone usually results in mutual accusations that one or other of the parties has pocketed it.

The grasslands of Africa and the jungles of India and Malaya are far from remote today, thanks to modern techniques of com-

Boas little danger to man

munication. If a human being were seized and devoured by a big snake, we can rest assured that such a gruesome and titillating incident would at once go the rounds of the world's press. Never having read of any such cases in recent decades, we can assume that they never or almost never occur. Big as they are, therefore, boas and pythons present little danger to man.

5. *The ostrich*

The assumption that animals are without rights and the illusion that our treatment of them has no moral significance is a positively outrageous example of Western crudity and barbarity.

ARTHUR SCHOPENHAUER (1788-1860)

No one who gives his girl-friend a hand-bag made of ostrich-skin need suffer a bad conscience because he is contributing to the extinction of the world's largest bird. Most of the leather used today comes, not from wild ostriches, but from South African ostrich farms, which have recovered from their total though temporary slump and are thriving once more. 42,000 ostriches are currently kept in paddocks there. Florida also has a few ostrich farms, though the big birds are maintained primarily for the benefit of tourists.

At about the turn of the century, ostrich breeding was one of the biggest businesses in South Africa. People were still paying anything up to £3,000 for a good stud-ostrich just before World War One. In those days, the demand was for feathers as opposed to leather. More than 3,600 tons of feathers were exported annually in the years round 1910, compared with only 1 ton seventy years earlier. Male ostriches have their plumage clipped close to the skin and are not plucked like chickens and geese.

An ostrich travelling at full speed can bound a good five feet into the air, which is why paddocks must be fenced to a height of six or seven feet. Aggressive male ostriches have to be handled with extreme care. One specimen at Hanover Zoo bent a centimetre-thick iron bar at right angles with a single kick. Another, this time at Frankfurt, ripped the clothes – underclothes included – off a

Nocturnal observation

keeper's back with one toe and half-hurled him through the wire fence at the same time. A grown man can ride a tame ostrich without imposing any undue strain on the bird.

Dr Klaus Immelmann, who wanted to study the sleeping habits of the ostrich, kept watch for several nights in the ostrich house at Frankfurt Zoo. He found that the birds spent between seven and nine hours every night sitting with eyes closed but head erect, a condition in which they are less readily disturbed by noise and movement than when awake. They get up at least a dozen times to urinate and defecate. In contrast to chickens and most other birds, ostriches excrete dung and urine in different places, although – like all birds – they do so through the same excrementory cavity. Ostriches which have been on their feet all day sometimes rest their heads wearily on top of a fence or prop them up in some way, shutting their eyes while still in a standing position.

What no one had realized before was that the ostrich also rests its head and neck on the ground while sleeping. This it does four times a night at most, and never for longer than sixteen minutes. Only then is the bird really soundly asleep to the extent that one can take flash-light photographs, stamp on the ground and raise one's voice without rousing it. Ostriches also like to extend their legs backwards on such occasions, these being normally drawn up beneath the body when the birds are seated. At no time did all the birds under observation fall into a deep sleep simultaneously. The same probably holds good for free-ranging ostriches, though nobody has ever observed the deep, genuine sleep of ostriches in the wild.

However, a quite similar posture can be observed on other occasions. A fleeing ostrich sometimes vanishes abruptly long before it reaches the horizon. Going in pursuit, one comes upon the bird sitting with its long neck laid flat along the ground. This is probably the origin of the legend which states that the ostrich hides its head in the sand in the belief that it cannot be seen. The Arabs were the first to write the story down, and it has since been perpetuated throughout the centuries by the Romans and their

The devoted father

literary successors. Half-grown ostriches are particularly inclined to adopt this position, but they quickly jump up and race off when approached.

In recent years, Drs Ingrid and Richard Faust have managed to hatch ostrich eggs in incubators at Frankfurt Zoo and rear the chicks without maternal aid. Ostrich chicks, which are already the size of a chicken at birth, require a great deal of attention from the outset. This is why so few ostriches have multiplied in zoos – a curious fact, considering the large numbers of birds bred by ostrich farmers. However, they raise ostriches under natural climatic conditions and allow the parents to hatch and rear their chicks in the normal way.

The male ostrich, a dutiful parent, scrapes a hollow in the ground and settles himself in it. Then, when the female has laid her eggs in front of him, he stows them beneath his body with neck and beak. A female living in the wild may lay as many as 8 eggs, and incubation takes about 40 days. In spite of high exterior temperatures, nest-temperature is maintained at 35-41.5°C. (95-106.5F.) Thus the eggs have to be kept cool, rather than warm, and protected from dehydration.

In Nairobi National Park a number of females jointly laid 42 eggs in front of a single male. Naturally enough, only 16 of them hatched out because the cock-bird was incapable of covering them properly. The male ostrich sits from late afternoon until early morning, so the female has far less brooding to do. Where other species are concerned, it does not matter so much if hunters concentrate on killing males. This does not apply to ostriches. An excess of eggs may prevent the surviving males from hatching out any eggs at all because they are incapable of leaving a proportion of the clutch to die and devoting their attention to the remainder.

An ostrich ménage at Nairobi National Park underwent a similar experience in 1960. The nest eventually contained over 40 eggs. What was more, the male ostrich had been foolish enough to site it so that visitors could spot it from the road. He and his harem were constantly surrounded by car-borne tourists who filmed and

Playing marbles with ostrich-eggs

photographed them from a range of two or three yards. Because the wild denizens of game reserves do not recognize human beings as enemies, the ostriches persevered. One day some young lions came across the clutch and played marbles with it, scattering the eggs over a wide area. Laboriously, the male ostrich raked them together into his hollow and continued to brood. Incredible as it may seem, some of the chicks actually hatched out. Even before hatching, chicks establish vocal contact with their parents and each other from inside the egg.

A male ostrich can roar like a lion. This it does by expelling air from the wind-pipe into the mouth, keeping the beak shut and thus forcing it back into the gorge or oesophagus, which dilates considerably. The cardia is simultaneously compressed to prevent air from entering the stomach. The whole of the bird's bare red neck swells like a balloon, and the result is a dull roar which carries for a long way. It was the ostrich's ill fortune to be endowed with magnificent plumage. The ancient Egyptians, who had noticed that the vexilla of ostrich-feathers are of exactly equal width on either side of the shaft, made them a symbol of equity and justice. All other birds' feathers have vanes of disparate width, so the shaft divides them 'inequitably'. The ancient Egyptians had also discovered that ostrich-feathers could embellish their personal appearance. As long as the feathers were only worn in the helmets of medieval knights, the hunting of wild ostriches was sufficient to meet the demand for them. Then, when it became fashionable to women to wear them during the last century, the birds' chances of survival began to look very slim indeed. Ostriches had long been extinct in North Africa and Egypt, and very few were still to be found in Persia and Arabia. The last ostrich vanished from Southern Arabia in about 1900. The last to be seen in the north of Saudi-Arabia was reported to have been shot on the Iraq border in 1933, though according to other accounts two ostriches were sighted and promptly killed in 1948 at the point where the borders of Iraq, Jordan and Saudi-Arabia coincide.

We owe it to the ostrich farms that these birds have not entirely

Ostrich-farms to the rescue

vanished from the face of the earth. Wild creatures which have been seized upon by fashion dwindle very rapidly as a rule. The scarcer they become, the higher the prices paid for their skins or whatever else they provide. In the end, because of their great rarity, prices soar to truly dizzy heights. The prospect of vast profits has prompted avaricious men to spend weeks, months or years in the most remote corners of the world in the hope of trapping the last wild mink, chinchilla or sable. Successful breeding of the same species in captivity enables skins to be offered for sale more cheaply. Prices fall, and it no longer pays to stalk or trap the wild specimens that still survive. Chinchilla, nutria, silver fox, mink and sable owe their continuing existence only to the fact that people have learnt, at the eleventh hour, how to keep and breed them in captivity. The first South African ostrich farm was opened in 1838. Thanks to the heavy demand for plumage, other farms were soon established in Algeria, Sicily and Florida – indeed, they were even to be found outside Nice in the South of France at the height of the ostrich-feather vogue.

Anyone holding an ostrich-egg in his hand for the first time is tempted to wonder how its inmate ever manages to emerge from such a prison without maternal assistance. The shell is as thick as a china cup – so thick that a mere human being has to use saw and hammer to break it. An ostrich-egg weighs between 3 lb. 5 oz. and 4 lb. 6 oz., or as much as two to three dozen hen's eggs. It is almost indistinguishable in flavour from a hen's egg, remains fresh and edible in a refrigerator for up to a year, and requires no special preparation. It is relatively simple to turn it into an omelette or scrambled egg. The best way to produce a nice big fried egg is to separate yolk from white and put the latter in the pan first, leaving space for some yolk with a circular strip of metal – a pastry-cutter, say. Then pour in the yolk when the white is half set. It takes nearly two hours to hard-boil an ostrich-egg.

Young ostriches hatch out after 42 days' incubation and then grow like weeds, averaging 0.4 inches per day. As soon as they can stand properly they start to perform the same frenzied dances as their elders. They make sudden dashes, gyrate, flap their wings, and flop down. Ostriches at Serengeti have sometimes been inspired by light aircraft flying overhead to put on wild cabaret shows of this kind. Young ostriches which have been artificially reared as pets in African homes follow human beings around like faithful dogs. If its foster-family goes bathing, the baby ostrich will boldly accompany them into the water like a duckling. At Serengeti, ostriches start to brood in September, and by Christmas they are already strutting about with their chicks.

Ostriches are excellent runners. These 8-foot giants can easily take strides of 12 feet when travelling at top speed. Driving behind them, one can tell by watching the speedometer that they are capable of maintaining 30 m.p.h. for periods of fifteen or even thirty minutes without showing signs of fatigue. Other wild creatures can travel equally fast, but only over short distances. Ostriches are said to attain a top speed of nearly 44 m.p.h. compared with the human sprinter's 19 m.p.h., so they must possess incredibly efficient hearts.

They can also be stout-hearted when circumstances demand it. Not long ago we encountered a male and female leading eight chicks. A hyena attacked them and tried to seize one of the young birds. There was a furious mêlée. The male concentrated on the chicks while the female went for the hyena, put it to flight, and pursued it for nearly a mile. A few days later we met the same family again. This time there were only six chicks to be seen.

The biologists Franz and Eleonore Sauer, both of the University of Florida, were the first to obtain detailed information about the life of wild ostriches in South-West Africa. Concealed in dummy termites' nests, some of them mobile, they managed to get quite close to the birds. Ostriches live not only in grassy plains but also in barren expanses of desert, areas of dense thorn-bush and even rugged highlands. They are doomed to die of thirst if permanently deprived of open water because aquiferous plants can only meet part of their water requirements. The birds manage to rear at least a few chicks at almost all seasons of the year. In addition to taking

A wife and two concubines

vegetable nourishment, they prey on lower forms of life and small vertebrates, many of which are captured after a zigzag chase.

Individual troops and families of ostriches frequently band together in peaceful communities which may number as many as six hundred birds, though the separate groups remain clearly discernible. Outsiders establish social contact by approaching in an 'appeasement' posture with tails pointed vertically downwards and heads inclined. One family will often adopt chicks or young ostriches belonging to another. Males sometimes join forces and roam about for days or weeks with 'nursery schools' of half-grown birds. Every troop of ostriches makes a hollow for the purpose of communal sand-bathing.

According to the Sauers' observations, ostriches live either monogamously or polygamously according to circumstances. Squabbles between males during pair-formation and the chasing of adult females often result in performances and 'dances' staged by whole troops. Normally speaking, each male has a principal mate and two 'concubines'. The senior female tolerates her juniors, and all lay their eggs in a communal nest. A nest found at Serengeti in 1968 contained no less than 45 eggs. The senior female, which has experience of hatching, generally chases the junior females from the nest as soon as their eggs are laid.

During pre-courtship, the male drives or lures selected females away from the troop by flapping his wings alternately, the senior female assisting him to round up yearlings that have attained sexual maturity. The new family then repairs to its breeding-ground, which is defended. During courtship proper, the male wanders off with one or other of the females and grazes with her for a while in a secluded spot. The two birds harmonize their movements with ever-increasing precision. It does not matter at all if no food is actually picked up, only that the birds' movements should be absolutely synchronized. Failure to achieve synchronization causes these preliminaries to be prematurely discontinued. If all goes well, however, the male becomes more and more excited. He flaps each wing in turn, flings himself to the ground and stirs up the dust

Fair game

with mighty wing-beats, twists and turns his neck in a series of rapid spiral movements. His dull roar rings out again and again while the female circles him in an appeasement posture with wings trailing. As soon as her mate jumps up she sinks to the ground and the male mounts her, flapping his wings.

No one had time to hunt ostriches in South Africa during the First World War. Ostrich-feathers having meanwhile gone out of fashion and become quite cheap, it was thought that the birds had grown too numerous to merit protection, and restrictions on their slaughter were eased. Smart operators chased them in cars and mowed them down wholesale, often returning from a single trip with four or five hundred skins which were subsequently turned into wallets and ladies' handbags. Nobody wanted the two or three hundredweight of meat on each carcase, especially when a bird might be nearly thirty years of age, so dead ostriches contaminated the whole area because hyenas and vultures were unable to cope with the sudden abundance of food.

Zoo staff who wish to handle an ostrich have only to pull a stocking over its head to render it defenceless and docile. Zoos where feeding is still permitted frequently run into trouble because of the incredible variety of articles which an ostrich will happily gulp down. Ostrich carcases have yielded coins, nails, half-horseshoes, pen-knives – indeed, one ostrich drank so much green oil-paint that its entire stomach and intestines were clotted with it.

Even in the latter half of the 19th century, Carl Hagenbeck was able to buy antelopes and ostriches at Suez which had actually been caught in the vicinity. In default of motor transport, most of these creatures had to be driven to the nearest railhead on foot, travelling in caravan. Incredible as it may seem to us today, Hagenbeck and his assistants simply haltered giraffes and led them along like horses. Once, when they tried to confine sixteen recently acquired ostriches in a paddock beside an inn, the birds broke out and ran off at lightning speed. One of Hagenbeck's men hit upon the idea of going after them with the mixed herd of goats, sheep and camels in whose company the ostriches had previously made the

Hagenbeck's trick

long trek to Suez. Sure enough, the agitated fugitives calmed down as soon as they caught sight of these animals. No sooner had the procession approached them than the ostriches reared their necks, beat their wings in delight, and danced round the herds of goats and dromedaries in a wide arc. Then, as if all were well again, the whole caravan set off for the railhead. The ostriches walked along between goats and dromedaries as quietly as if they were held there by an invisible force. They offered little or no resistance and allowed themselves to be led into the wagon reserved for them. It should, however, be noted that they had already spent no less than 42 days untethered while accompanying the goats and dromedaries from Kassala to Suakin.

6. Floating among crocodiles

When the great baobab-tree falls, little goats climb and caper upon its trunk.
AFRICAN PROVERB

Man's love of animals is by no means evenly and fairly distributed among all living creatures from the earth-worm to the elephant or the amoeba to the chimpanzee. Human beings are better disposed towards creatures they can cook and eat than towards those which are capable of devouring them. For instance, I have yet to come across many animal-lovers who enthuse about crocodiles, even though crocodiles were venerated as gods in ancient Egypt and elsewhere.

Crocodiles are not, it is true, quite as dangerous as adventure stories make out. Frank Poppleton, a game warden acquaintance of mine, used every morning to swim across the Victoria Nile, which is renowned for its large numbers of crocodiles. He contracted bilharzia, a kidney disease caused by tiny worms, but never had any trouble with crocodiles himself although one of his African assistants sustained a lacerated leg. Nevertheless, in the days when crocodiles still abounded in Africa, it did occasionally happen that human beings were killed by them.

In September 1962 some children belonging to a settler named William Cox, three boys ranging in age from eight to twelve, were bathing in the Kabue at Chingola. John Maxwell, a British policeman who was also bathing in the river, caught sight of a 16-foot crocodile swimming towards the children. He immediately dived and hoisted the boys on to a rock but was seized by the leg and dragged under. Being a resourceful and athletic man, he com-

A brave young African girl

pelled the crocodile to release him by gouging out its eyes. Meanwhile, a young African woman named Malomi had come running in response to his cries. Although unable to swim she plunged into the water, helped Maxwell on to her back, and carried the gravely injured man to safety on all fours. John Maxwell had his left leg amputated, was flown to England and invested with the George Medal. Africans often display incredible courage and readiness to help on such occasions. Adventure stories and books for the young have a regrettable tendency to ascribe qualities such as dash and daring to American Indians rather than Africans, and this easily creates a false impression.

A few years ago I myself had to stitch up the hand of a friend of mine, the biologist Ian Parker, using his wife's sewing kit, because he had been bitten by a medium-sized crocodile lying in the shallows. I must confess that I do not personally enjoy bathing in crocodile-infested waters. Apart from anything else, inability to see the reptiles does not preclude their presence. A full-grown crocodile can remain submerged for more than an hour without breathing.

The fact that one finds a creature unattractive does not rob it of interest. Tourists visiting the national parks of Africa are keen to see crocodiles even if their flesh crawls in the process. However, crocodiles are virtually extinct everywhere in the world. This is because their skin has become fashionable for conversion into wallets, handbags and women's shoes, and each square inch of crocodile skin fetches a good price. 12,509 skins were exported from Tanganyika as long ago as 1952, and crocodile was not as fashionable then as it is now. Even in the United States, game wardens and policemen in Florida are waging a hitherto futile campaign against organized gangs of poachers who raid protected areas for their surviving stocks of alligators – which are, incidentally, harmless to man. Those arrested have even included corpulent ladies who turn out to be wearing bloody, freshly stripped alligator-skins under their dresses.

The finest and oldest crocodiles in Africa live at Murchison

Contemporaries of the early explorers

Falls National Park in Uganda. Travelling up the broad, fast-flowing Victoria Nile in one of the motor-boats from the tourist hotel, one can see dozens of the old gentlemen sunning themselves beside the river or on sand-banks. They have long grown accustomed to the boatloads of visitors, just like the local Cape buffalo, elephants, hippos and rhinos. It is the only place in Africa, if not the world, where one can be reasonably certain of finding crocodiles to photograph.

Years ago I lay at anchor here for hours in a small canoe, watching the huge saurians and tanning nicely in the near vertical rays of the equatorial sun. I squinted across at the ancient grey-green monsters, which were doing just as I was – sunning themselves and eyeing me lazily in return. Every now and then came the splash of a big fish, and small clumps of green Nile cabbage drifted swiftly past. It was easy to grow idle and forget about my camera – after all, the old gentlemen across the way weren't exactly *doing* anything. Some of these old crocodiles had been lying in their eggs half a yard underground before the first white man came up the Nile. It was possible that they had stared quite as indifferently at the British explorer Samuel Baker as he came paddling up the river eighty years before accompanied by his pretty young Hungarian wife in long Victorian skirts.

The crocodile's annual rate of growth in the first 7 years is $10\frac{1}{2}$ inches, but once they have attained the age of 20 they grow only 1.4 inches a year; so those which measure over 18 feet were probably born more than a century ago. The biggest ever shot in this area was over 20 feet long, or too big to fit into the average living-room. The stomach of another big crocodile killed in Uganda not long ago yielded a python more than 15 feet long. Two years ago one of our game wardens saw a group of five drinking lions take fright at something invisible to those who were watching. At that moment a young lion was seized by the leg, dragged into the water and drowned.

Crocodiles are modest trenchermen, even so. Their quiet way of life requires little expenditure of energy. Having examined the stomachs of 263 specimens ranging in length from 6 to 14½ feet, the biologist H. B. Cott found that 54.8 per cent contained either no food at all or only indigestible remains such as hair, scales, claws, etc. Only 68 per cent of the creatures had made a good meal just prior to death. At our zoo, a full-grown crocodile takes 150 days to consume its own weight in food. A pelican ingests one-third of its body-weight in fish at a single meal, so crocodiles cannot truly be described as voracious.

The Murchison Falls National Park in Uganda is such a crocodile paradise that zoologists wishing to study the living habits of these big saurians have always gone there.

Such, at least, was the case until a few years ago. There are still crocodiles in the Victoria Nile, it is true, but they are growing scarcer month by month. Lured by high prices, natives from the fishing villages along Lake Albert row into the Nile under cover of darkness, dazzle the creatures with spotlights and shoot them between the eyes or spear them. Game wardens have been stationed on the bank, but how can they reach the water if they see dark shapes come gliding down the river? A game warden parked high above the Nile in his field-car has to drive ten or twenty miles, board a motor-boat and chug back up the river again. By the time he gets there the poachers have long since vanished into the bush and their boats are probably concealed in readiness for the next foray.

'Too bad we can't drive our cars through water and flit through the bush in a motor-boat,' one game warden said to me.

I recalled this remark when I read somewhere that a German firm was building cars capable of travelling on land and through water. It occurred to me that a car of this type would be ideal for waging war on the water-borne poachers of Uganda. It could cruise along the bank at speed and simply take to the water as soon as suspicious shapes were seen on the river. The poachers would have no time to do a disappearing-act. One could track them with the car's headlamps, and there would be a prospect of pursuing them even if they abandoned their boats and made off overland.

Screws at the rear

The idea was certainly worth a try, if only because it might put new heart into the game wardens engaged in this unending struggle.

I duly ordered one of these amphibious cars and had it shipped straight to the Kenyan port of Mombasa. My associate Alan Root took delivery of it there and drove the 1,300-odd miles to Uganda. When I landed at Entebbe (after a nine-hour flight from Frankfurt) he was waiting for me with a smart red car. I stowed my bags inside, folded the roof down, and drove to the national parks.

The car hummed along at seventy, looking more like a handsome modern convertible than a cross-country vehicle. Whenever we stopped for petrol, however, we were surrounded by a crowd of squatting, peering Africans who had quickly noted that the back of the car was adorned with two propellers. The idea of a car which could travel through water astounded them, nor were they incredulous when I flippantly assured them that it could fly through the air as well.

There was, we discovered, no need to drive slowly and gingerly into Kazinga Channel in Queen Elizabeth National Park or into neighbouring Lake Edward, across which lies the Congo. We soon grew accustomed to driving in briskly, hitting the surface with a splash which sent water cascading over the windscreen. One movement of a lever was enough to transfer the drive from roadwheels to propellers, and we continued to steer the car in the normal way because the front wheels deputized for a rudder. It was fun to watch the astonished faces of a ferry-load of Africans as we overhauled and circled them. The car's two doors extended below the surface but were completely watertight when clamped shut by means of a special lever before the vehicle left dry land.

Needless to say, a water-car of this type makes a perfect toy for grown-ups. A hippo family at first took my Amphicar for a boat. A bull surged menacingly towards me for a few yards, but when I continued to drive on, undeterred, the twenty big heads submerged silently well before I reached them. I could detect the animals' whereabouts by the trails of air-bubbles which streamed outwards in

Dunked by a rhino

all directions. Their broad feet were releasing gases from the mud as they paddled along the bottom.

One bull did not take flight. All at once I felt the car's wheels pass over something round – something live. The entire vehicle gave a sudden upward lurch, but it sat so securely in the water that I never for a moment entertained the possibility of being capsized. Two years before, over on the Congolese side of Lake Edward, a female hippopotamus had attacked a large metal-hulled boat of mine. Although the boat did not capsize, an African perched on the gunwale sailed through the air and landed in the water, lacerating his behind on one of the animal's tusks. He suffered no further injury. The fat bull could not easily have catapulted me out of my amphibious car because I was comfortably ensconced behind the wheel on an upholstered seat, in a perfect position to observe 'the conquest of Nature by human technology'. I should only have come to grief if the bull had punched a hole in the car's body-work with its tusks and so sent the whole contraption to the bottom. Fortunately, the animal was not in real earnest.

Alan Root had actually filmed my car being hoisted by the hippo whose presence I had not allowed for. He was in far more danger than I, being seated with his wife, Joan, in a cockleshell of a boat equipped with an outboard. The boat's transom cleared the water by little more than an inch, and Joan had to bail continuously. One nudge from a hippopotamus and they would both have been in the water, so they kept their distance and used a telephoto lens.

Elephants seldom if ever pursue an adversary into the water, so I boldly steered for a group of tuskers drinking at the water's edge. Although I switched off in good time and drifted silently towards them, the four huge creatures became restive. They shuffled backwards in the soft sand, flapped their big ears agitatedly back and forth, scrambled up the steep embankment behind them, and disappeared. All was quiet again, and no one would have believed that the grey giants had been there a few moments before.

One big lone bull elephant behind the next spit of land was com-

Putting an elephant to flight

pelled to take a closer interest in me whether he liked it or not. He had been careless enough to walk along the flat narrow strip of shore, close to the water, and the bank on the landward side was too smooth and sheer even for a climber of the elephant's agility. Consequently, he could not slink away. Chary of retracing his steps along the water's edge because my car and I had by this time drifted close in-shore, he retreated gradually between water and bank until his route was barred, after thirty yards or so, by a large tree. With that, he was trapped. Being unable to run away, he took a few quick steps towards me, trumpeting and splashing through the shallows. Then, when the red thing in the water did not take flight as he had expected, his courage again deserted him and he backed off. I might add that it would have taken me a moment or two to start up and put the car into reverse. In the end, overcome by desperation, the ponderous old gentleman decided to make a dash for it along the shore. He charged past me with tail erect, then streaked up the bank at a more accessible spot.

It is easy to feel superior when a huge animal turns tail and runs away. I reminded myself how much humbler I felt when proceeding through the African bush on foot. All the same, it was a 'new experience in motoring' and I naturally relished it as I sat there in my comfortably upholstered seat. Caution was indicated only where the shore shelved very gradually, in case I got stuck at a highly inopportune moment. There was little fear of damage to the screws because they nestled high up between the rear wheels and were thus prevented from fouling the bottom. Even if one ran aground, it was relatively easy to extricate the car by rotating the rear wheels. On the other hand, it did take a little while to change gear and get under way again, and this could be slightly unnerving when an infuriated Cape buffalo was making threatening lunges a few feet away on shore. If water slopped into the vehicle on such occasions one had only to operate a knob and it was ejected by an electric pump.

I read in the paper that two people were drowned in a similar vehicle at Hamburg not long ago, quite why I do not know. They had probably put the hood up and could not get out in the general excitement. I deliberately drove with the hood down, which was another new experience. In Africa we normally use field-cars which enclose their occupants like boxes unless they care to climb on to the roof through a manhole. This car was far better adapted to photography. Prides of lions had no objection when I visited them and drove cautiously between lionesses and their young.

The lions behaved just like all the lions in our modern national parks. That is to say, they cut presumptuous car-borne visitors down to size. Not that they attacked or even snarled at us – far from it. They simply declined to notice us. If lions are lying in the shade, no field-car or gaggle of tourists is going to budge them three feet into the sun for the sake of a better photograph. A lioness may be looking straight past you. Cough, shout, stand on your head – nothing will persuade her to favour you with a single glance. On the other hand, if another lion looks at you because it happens to suit him, you get the feeling that he is looking straight through you – that you are just so much air to him.

Not to be noticed is something which most people find extremely hard to endure. They feel a temptation to pelt the King of the Beasts with stones in the hope that he will at least favour them with a fleeting glance. What usually happens is that the big yellow cat rises to its feet, walks obliquely towards them, passing within eighteen inches of their car's rear wheels as though the vehicle did not exist, and vanishes behind the nearest hummock.

The lions which I eventually discovered beside a fallen euphorbia-tree, surrounded by very barren terrain, were equally unimpressed by my red car, even though they had probably never seen such a brightly coloured, open-topped vehicle before. There were three big lionesses with half-grown young, also a cub of perhaps four months. At least the baby showed some interest in me. It toddled away from the three lionesses, eyed me briefly, and circled the red Amphicar. It was now out of sight of its mother. Would her suspicions be alerted, would she become nervous and approach me? Not a bit of it. I might not have existed, though she was so

The incautious fisherman

close that I could have counted her whiskers. And yet, one of the animals facing me had committed a heinous crime.

Precisely one night earlier, for the first time in fifty years, a man had actually been devoured not far from there on the borders of the national park. He had reached the Congolese frontier-post late that evening on his return from a fishing-trip. The customs officials refused to let him through and told him to spend the night at the nearest house, a hut beside the road. The occupants of the hut raised no objection until he insisted on taking his catch of

Page 117:
My amphibious car coped safely even with the seething, swirling water below the huge falls. The Murchison Falls are 130 feet high and form the centre-piece of the national park named after them, whose 12,000 square miles make it one of the largest in Africa. President Obote of Uganda plans to make this major tourist attraction the site of a huge generating station which will utilize 90 per cent of its water below ground.

Page 118, above:
This crater is situated in the Ngorongoro game reserve. The Ngorongoro Crater itself has become a favourite haunt of tourists. No roads yet lead to the Embakai Crater, which makes it extremely inaccessible. Would-be visitors have to camp en route, although the night temperatures are unpleasantly low at this altitude. The flamingos feed in the crater but prefer to breed in the shallow, salty waters of Lakes Natron and Magadi.

Page 118, below:
I drove my amphibious car straight into herds of hippopotami. They dived at the last moment, but on two occasions I was lifted by bull hippos. The car withstood this treatment without capsizing. I should only have been in trouble if the animals had decided to punch holes in the thin body-work with their tusks.

Page 119, above:
Pelicans devour very large fish, for which purpose they have an unusually expansible crop. One can easily thrust one's fist and forearm down a pelican's throat as far as the stomach. We once retrieved a missing bunch of keys in this way at Frankfurt Zoo. At Rome Zoo, coots and dabchicks often used to fall into the pool in autumn. One day a pelican swallowed a dabchick alive. The little bird must have clawed and pecked away busily at its host's insides, because it was very soon returned to the water unscathed. The pelican never touched another dabchick or coot. Here, pelicans are flying back through the twilight from Lake Manyara to their breeding colony and perches among the trees.

Page 119, below:
Elephants and white rhinos get on well with other animals. Generally speaking, the elephant is the real King of the Beasts. All other animals give way to it, lions and rhinos included. Animals which have to live at close quarters often coexist peacefully and may even make friends, as here, in the unusual new zoo at Boras in Sweden.

Nothing left but a hand and a foot

dead fish into the hut with him. They balked at this, so he slept on the threshold with his fish beside him.

The only immediate sign of him next morning was a trail of blood leading up a small incline and into the bush, where a large pool of blood was found. The lion had left nothing but a hand and a foot.

In defence of lions I should at once add that many more people have been killed in recent decades by cars on the road leading through Queen Elizabeth National Park. A few days earlier a truck had ploughed into a group of lions, killing one and injuring the rest so badly that they were left hobbling.

I could have sold my amphibious car three times over to enthusiastic American tourists, probably at a premium, but I and Aubrey Buxton had already made a joint present of it to the Uganda National Parks. It is to be hoped that their new African director, Francis Katete, will put paid to the poachers with its assistance. The Zoological Society of Frankfurt has also presented him with a motor-boat for combating game poachers on the Nile.

A few months later I received the following letter from Alan Root:

Dear Bernhard,

I am now out of hospital at last but am still very weak and my arm and hand are still swollen and painful. It happened at Joy Adamson's camp in the Meru Park. I found the snake, about 4 feet long, near her camp and caught it to show it to her and an American that we had with us. I opened its mouth and milked it a bit to show them the amount of venom that these snakes have, and then I put it down whilst she was changing film in her camera. I then went to catch it again but the snake was by now very angry and as I grabbed for its neck it turned and bit me. Because I was bitten 10

Page 120:
The South African gemsbok is the most conspicuously marked of the four subspecies of oryx. It occurs in exceptionally arid regions, like the Kalahari Desert, where it uses its front hoofs to dig up the chamma, a water-bearing fruit resembling a melon, complete with roots.

A serious snake bite

years ago and had injections of antivenin I knew that I would be sensitive to further antivenin, and therefore I delayed the injections for a while in order to see how severe the bite was going to be. Because I had already partly milked the snake I did not think that I had got a very big dose of poison. However, the arm swelled rapidly, I became dizzy and nauseated, so Joan gave me the 3 injections (3 ccs.) intramuscular whilst flying back to Nairobi.

As we expected, I reacted violently to the antivenin, vomiting and my heart-beat and respiration going up rapidly. When we got to the hospital in Nairobi the doctor thought that these conditions were due to the snake bite and gave me another 10 ccs. of antivenin intravenously. I immediately went into anaphylactic shock and they had quite a battle with antihistamines, cortisone, adrenalin and oxygen to keep me alive.

Over the next 3 days my condition deteriorated. The hand was three times normal size and covered with enormous black, blood-filled blisters. My arm swelled to the size of my leg and was also full of blood, and I had great bloody swellings stretching from my arm pit down to my waist, across my shoulder blades and in front as far as my throat. On the fourth day my haemoglobin count had gone down to 36. 40 is considered to be the lowest permissible level. I was given a transfusion of 4 pints of blood which made me feel a lot better and from that day on I started to improve.

Joan got Prof. David Chapman, an expert on snake bite treatment, to fly up from South Africa to see me and his advice was of great value. During these first 4 days the doctors had been seriously considering amputating my arm for they had not been able to feel a pulse at all in the arm, and the hand and fingers were so swollen and damaged that they considered it would never be of any use to me again. However, now, 4 weeks after being bitten, the arm is almost back to its normal size and my hand has made a miraculous recovery. It is still very swollen and painful, but my thumb and 3 outer fingers are all in reasonably good shape. I can move them and I have most of the sensation back now.

My index finger, which one fang obviously went deeply into, is

The severed forefinger

not so good. The whole top surface was destroyed by the venom right down to the tendons. Within the next couple of weeks I will be having new skin grafted on to this finger and it will then be a matter of time before we know whether it is going to be of any use to me. You told me about your finger that the chimp bit and I have noticed that you are able to use that hand very well despite it being stiff. It will obviously be anything up to 2 months before I will be able to do anything with my hand again, and though I am impatient already I really should be grateful that I have got away so lightly.

     Best wishes from us both.
      Sincerely,
       Alan.

Alan's index finger had to be amputated after all. His thumb retained only limited mobility, so he flew back to England for another two months and underwent renewed and highly successful surgery on his hand. When he returned, all his friends commiserated with him on his loss. His response was to reach into his pocket and proffer a severed, gory index finger which he had supposedly preserved for sentimental reasons. In fact, he had bought himself a fake finger in soft plastic from a novelty shop in London.

Shortly afterwards I was visiting the Okavengo Swamp in Botswana (formerly Bechuanaland). My host's father-in-law, Mr Robert Langley-Elton, aged 51, had been bitten in the calf by a poisonous mamba while crocodile-hunting exactly four weeks earlier, on December 19th 1968. The incident occurred more than two hundred miles from his home, just as he was getting into a boat. He injected himself with serum half an hour later, but smashed one of the two ampoules in his agitation. He was unwise enough to try to drive home and died on the way, four hours after being bitten. He lies buried on an island near the camp.

So instances of snake-bite do sometimes occur in Africa. However, the ordinary tourist has little prospect of even seeing a snake unless he comes across one lying dead in the road.

7. Locusts on the march

He who beats a dog beats its master.

AFRICAN PROVERB

'And the Lord said unto Moses, Stretch out thine hand over the land of Egypt for the locusts, that they may come up upon the land of Egypt, and eat every herb of the land, even all that the hail hath left.

'And Moses stretched forth his rod over the land of Egypt, and the Lord brought an east wind upon the land all that day, and all that night; and when it was morning, the east wind brought the locusts. And the locusts went up over all the land of Egypt, and rested in all the coasts of Egypt: very grievous were they; before them there were no such locusts as they, neither after them shall be such. For they covered the face of the whole earth, so that the land was darkened; and they did eat every herb of the land, and all the fruit of the trees which the hail had left: and there remained not any green thing in the trees, or in the herbs of the field, through all the land of Egypt.'

Modern Europeans are tempted to dismiss this Old Testament plague as a myth – quite wrongly so, because it still devastates wide areas of the world. As recently as 1873, 1874 and 1875, swarms of locusts penetrated Central Germany, most of them from the regions around the Black and Caspian Seas. The four-winged armies hummed across Poland and Galicia, reached Silesia and Brandenburg, and pressed on into France and the British Isles. In Brandenburg, 4,425 German bushels of locust-eggs were collected from less than 5,000 acres of land, but ten times that quantity was collected during a locust campaign in Cyprus. Breslau and Gotha struck coins in commemoration of the locust invasion. In 1879-80 people in the south of Russia had to shut their houses to keep out the swarms of insects, and streets became impassable. At Elisabethpol the water had to be filtered because canals and watercourses were clogged with locusts. No bread could be baked because even the bake-ovens were choked with masses of dead insects. Trains stopped running in the Don Steppe because locusts on the track made their wheels spin like soft soap.

In 1955 a swarm of locusts 150 miles long and 12 miles wide descended on Southern Morocco. In 1961 and 1962 the combating of locusts there was rendered more difficult because heavy rains had closed the muddy roads to motor traffic. Within five days the locusts inflicted damage to the tune of one billion old francs on an area of more than 2,000 square miles. In the Sous Valley more than one-fifth of the land under cultivation was wholly destroyed and the remainder almost so. In five days the locusts devoured 7,000 tons of oranges, or nearly 60 tons an hour – a quantity in excess of total French consumption. It took about five years to repair the damage, though more swarms arrived in the interim, some of them capable of ravaging more than 60 square miles at a time. Over two dozen aircraft were employed to fight them with insecticides. Without this modern form of pest-control and weather favourable to its use, the big citrus plantations between the Atlas and Anti-Atlas would have been destroyed.

South African trains ran late in 1966 because the tracks were covered with locusts. Hundreds of cars transported insecticide in order to protect the farming areas of the Orange Free State. This campaign cost more than half a million pounds.

Of the ten thousand or more species of locusts in existence, only about five undertake such mass migrations, notably Schistocerca gregaria, Locusta migratoria, Schistocerca paranensis, Locustana pardalina, and Nomadacris septemfasciata.

All the many other hopping and chirping locusts live singly and avoid coming into too close contact except when mating. Migra-

Early successes in the war against locusts

tory locusts, by contrast, make a habit of close contact. The migratory urge comes upon them only occasionally and only in areas particularly favourable to migration. The voracious hordes usually emanate from these places alone. Recognition of this fact was fundamental to the many successes that have since been gained in the campaign against the insect armies.

Locusts do not become sexually mature for some weeks or months after they have moulted for the last time and acquired wings. Most species also change colour at the same time, females becoming straw-coloured and males lemon-yellow. Members of many species recognize each other by the sound of chirping alone. The female seeks the male only so long as he continues to chirp; as soon as he stops she wanders about aimlessly. Migratory locusts congregate in such numbers that the sexes have little difficulty in finding one another, but the pitch and quality of chirping prevent members of different species from mating. Each locust copulates several times.

After the females have been mounted by the males they begin to deposit their eggs in the ground. For this purpose they have four chitinous plates at the extremity of the abdomen, which is considerably elongated by the extension of intersegmental annular membranes. Having used these plates to dig a hole in the sand, they lay between thirty and a hundred eggs in it and plug the chamber with a frothy secretion which admits air. During a locust invasion of Algeria in 1890, an estimated 2,720 billion locusts were destroyed in a limited area, including about 560 billion eggs, 1,450 newly hatched young, and the eggs in females killed at the same time. A female buries ten such batches after one fertilization. Areas in which these vast numbers of budding locusts are concealed can readily be identified by the white blobs of froth. Fifty to sixty days later, depending on warmth and humidity, the ground is covered with myriads of tiny insects which crawl along like worms. Approximately 8 mm. long, they have to extricate themselves from a sheath before acquiring full mobility but at once set off en masse in a specific direction, crossing hills, ditches and even rivers at the cost of enormous casualties. During their earth-bound phase the larvae devour anything they can find, increase in size, and shed another five skins. At the end of that time they are over 3 centimetres long and already have wing-stumps, but are as yet unable to fly.

In order to moult for the last time, they seek a firm perch on a twig, still eating incessantly, and become motionless. The new skin, or future chitinous casing, once again lies crumpled and folded beneath the old skin, which has become too tight. An accumulation of blood and air forms on the creatures' backs, causing the old integument to split lengthwise. Slowly, very slowly, the locust crawls out in its final form. The wings are still folded in an arc on its back. Gradually, under the pressure of blood streaming into the numerous blood-vessels, they unfold and become smooth. The locust, fully developed at last, beats its wings. At least three weeks and possibly several months must elapse before it is capable of breeding.

The ancient Greek historian Herodotus reported 2,400 years ago that the inhabitants of the desert used to gather locusts and dry and grind them into a powder which they mixed with milk and ate. Alfred Brehm wrote that locusts tasted disgusting and had little nutritive value, but visitors to the oases of the Sahara are still offered, apart from dried dates, locusts crisply fried in oil. To eat them, one first breaks off the head, wings and lower parts of the legs. They taste far from revolting. Not only are they nourishing and rich in fat, but their intestines are crammed with vegetable matter which undoubtedly has a high vitamin content. John the Baptist, who spent forty days in the wilderness on a diet of locusts and honey, was not, therefore, being unduly ascetic. Even today, the desert tribesmen known as Tuaregs use flour made from locusts as their sole form of nourishment while travelling in caravan. Its unpleasant odour disappears when it is mixed with milk.

According to Franz Kollmannsperger, locusts need no drinking-water in the southern Sahara. They must ingest food continuously, on the other hand, and cannot go hungry for any length of time.

A personal refrigeration plant

Their digestion is equipped to extract moisture from dry vegetable matter, hence their habit of devouring any plants, however dehydrated. Unlike that of many beetles, their chitinous casing is quite pervious to water and does not hinder evaporation. This comparatively high rate of evaporation lowers their body temperature and endows them with a cooler 'physical climate'. The insects are therefore resistant to heat and will even land in areas where the ground temperature is 58°C. (136F.) According to R. Chapman, migratory locust larvae (Schistocerca gregaria) in the second stage of development sought to avoid degrees of warmth below 32°C. (89.5F.) and above 43°C. (109F.) in the temperature-organ. (A temperature-organ is a kind of natural thermostat which enables creatures to select at will from a gradually rising scale of temperatures.) All the migratory locust species tested by him preferred temperatures in excess of 40°C. (104F.)

Many areas, especially warm and humid ones with plenty of vegetation, are veritable breeding-grounds for armies of migratory locusts. This is true, for example, of the Rukwa Valley in the southwest of Tanzania, part of the Great African Rift. Hordes of them have frequently flown to South Africa from there. That is why about a dozen European colonial officers were stationed in this remote valley as long ago as 1950 and provided with houses, roads, bridges, and landing-strips for aircraft. It is naturally advisable to treat locusts with insecticide as soon as they emerge from the egg. According to Chapman's observations, the Red Migratory Locust (Nomadacris septemfasciata Serville) never takes to the air at temperatures below 18°C. (64F.) and seldom below 22°C. (71.5F.), its preference being for temperatures between 29°C. (84F.) and 35°C. (95F.) Little flying was done in winds exceeding 3 metres per second. The locusts tended to fly into light winds and with strong winds, so sections of the same swarm might be flying in opposite directions at different altitudes.

After moulting for the last time and releasing their wings, locusts begin by making circular flights and returning to base. They do not fly off in earnest until most of the swarm has assembled, although they will do this even when their native area still has quite enough fresh vegetation to offer.

One is naturally led to ask what impels species of migratory locusts which normally avoid each other to form such immense swarms and fly off into the blue. Experiments by Faure, Husain and Mathur point to overcrowding as the cause. The more often locusts encounter each other, the more they cease to be solitary creatures and become soldiers in a huge army. The scientists accordingly kept numerous young larvae belonging to non-gregarious species of migratory locusts crowded together in a container, and they turned into mass-migrants even there. This does not occur suddenly, however, even in the wild. One generation must always intervene in which the number of moults becomes fewer (only six instead of seven), the difference between male and female less conspicuous, and the larvae darker. It is from the eggs of this generation that there emerge young with the urge to migrate. One generation must always intervene even when mass-migrants revert to being locusts which live singly. Young larvae of the migratory type flock together whereas the solitary insect tends to diverge.

It is evident that these migrations are not aimless. The locust armies which reached Germany and England had probably lost their way, but the locusts which fly across the Sahara breed in Morocco from March to July and travel south again to lay more eggs in the plains of the Southern Sahara and the Niger area between July and October.

8. The hyena changes its image

Were the sun to rise at midnight one would find that not only the hyena is evil.
AFRICAN PROVERB

Hyenas can be quite nice animals in my experience – in fact I have known some extremely likeable specimens. They become quite tame and friendly if one takes trouble with them. However, I have always been aware that I am virtually alone in my views. Ludwig Heck wrote in the last edition of *Brehms Tierleben* that they are 'ugly, misshapen and clumsy devourers of carrion and bones'. The eating of carrion does not in itself merit our abomination. I am just as much a carrion-eater as you are, nor do we generally kill the animals whose meat we consume. Few of us would be capable of doing so.

The popular theory, often shared by African hunters and farmers of long experience, is that hyenas wait until lions, leopards and other courageous predators have killed and then either try to filch a piece of meat or devour what remains after the so-called king of the beasts has eaten his fill. Feeding lions are, it is true, often seen surrounded by hyenas, jackals and vultures which wait hungrily and patiently until they have finished their meal.

I and some of the zoologists working at Serengeti doubted occasionally whether the current picture of the hyena – a picture consistently reproduced by every book written during the past century-and-a-half – was really accurate. For that reason, my friend Alan Root and I recorded lion- and hyena-calls on tape and played them through an amplifier at selected spots. A most surprising fact emerged. Hyenas took no notice of the roar of lions, whereas lions were attracted by the yelping 'laughter' often emitted by hyenas when a pack of those animals is greedily and hurriedly dismembering and devouring prey and squabbling at the same time. Few lions can resist the sound. One after the other, they abandon the dolce far niente which is their usual pastime and make for the source of the hyenas' cries. By broadcasting the call of the hyena we persuaded whole prides of lions to congregate round our Volkswagen bus – indeed, lured them to the summit of hills where they would never normally have gone because rhinos, antelopes and giraffes never go there either. Hence the notice-board on a hill overlooking part of the Amboseli reserve, which reads: 'Exercise caution – You may get out of your car at this point only'. We lured lions to the very foot of the post bearing this reassuring sign by playing taped hyena-calls. Then, for fun, we photographed notice-board and lions together.

When a hyena feels lonely and wishes to make contact with other members of its pack, it inclines its head slightly towards the ground and emits a long-drawn-out but not unduly loud howl. We recorded this call and sent it echoing among the rocks, whereupon four or five hyenas hurried to the spot and gathered round the loudspeaker looking puzzled and expectant.

If hyenas have killed an animal themselves and utter their excited giggling cries, it is seldom long before some lions amble up, chase the hyenas away, and treat themselves to a free meal. The converse is far rarer. Some hyenas recently succeeded in driving lions off their kill at Mikumi National Park in Tanzania, but in this case the eleven lions included eight young. The hyenas, also eleven in number, crept up under cover of the tall grass and made the lions increasingly nervous and unsure of themselves. Normally, hyenas stand around waiting. Most animals are hunted and killed at night. Coming upon a dead zebra by day, we automatically assume that it was killed by lions and that the hyenas are merely carrion-eaters-in-waiting. Our experiments with tape-recordings suggested otherwise, but that was only conjecture. To zoologists, only firm evidence counts.

Fifty doped hyenas

This evidence was recently adduced by the young Dutchman Dr Hans Kruuk while studying hyenas at Serengeti and in the celebrated Ngorongoro Crater. We shall in future have to regard these animals in quite another light.

First, Dr Kruuk and his wife had to sacrifice a lot of sleep, especially on bright moonlit nights. All previous judgements had been formed on the basis of observations made during the daytime, yet the spotted hyena is mainly active at night. The huge Ngorongoro Crater lends itself particularly well to such observations. The same animals reside there permanently, and seldom if ever scale the 1,500-foot-high sides of the crater to reach the outside world. Because no one has done any hunting there for so long, the 4,000-odd zebras, 10,000 gnus and numerous other animals have become increasingly tame and take little notice of cars or human onlookers. Ngorongoro is a vast zoo full of wild animals living at liberty in their natural state. Animals in the broad adjoining plains of Serengeti do not have quite the same faith in human beings because they make long migrations and encounter poachers and hunters outside the national park at many seasons of the year.

Roughly four hundred spotted hyenas live in the Ngorongoro Crater. Hans Kruuk succeeded in knocking out fifty of these with anæsthetic darts and attaching ear-tags to them so that he could recognize every individual hyena on future occasions. He sighted forty-five of them again within the next six months, almost all in the district where they had been tagged. There are eight large hyena territories on the floor of the crater, which covers approximately 100 square miles. Each territory is inhabited by a pack which may number between eighty and a hundred head. Of the tagged hyenas, only the males were seen outside their territory, never the females. The five tag-wearers which were never sighted again were all males and young. Hyenas generally hunt within their own territory. If the quarry runs into a neighbouring territory the hunters are often driven off by other hyenas. Generally speaking, spotted hyenas which come from elsewhere and do not belong to the resident pack are liable to attack and hostile treat-

Trekking 50 miles for a meal

ment. The pack has dens in its own hunting-ground – very often a maze of holes and subterranean passages concentrated in a single area.

Things are somewhat different in the open Serengeti. Here, the Kruuks have so far tagged a hundred hyenas. Although hyena packs still have their own territories and fixed abodes in the broad plains roamed by large herds of gnus, zebras and Thomson's gazelles, they spend much of their time trailing the herds as they graze. They sometimes remain away from their den for several days, and may be encountered up to fifty miles away. A game warden at the Kalahari National Park in South Africa had some goats carried off by hyenas. He was so furious that he followed their tracks. Their den turned out to be twenty-five miles away, so the hyenas had trekked for fifty miles in order to steal his goats. The South African authority F. C. Eloff likewise stated in 1964 that hyenas kill their own prey and seldom eat carrion. Of 1,052 hyenas observed while taking food, no less than 82 per cent were devouring animals killed by their own kind and only 11 per cent the prey of other creatures such as jackals, lions, wild dogs, leopards and cheetahs. In the remaining instances the identity of the killer was uncertain. More than 30 per cent of the hyenas encountered while eating in daylight were feeding on other predators' kills – hence, presumably, their reputation for eating carrion.

African spotted hyenas are capable of indulging in the strangest behaviour. Mervyn Cowie, who has since retired from the directorship of Kenya's national parks, told me that whole herds of hyenas used to live off the refuse from the Mbagathy slaughterhouse outside Nairobi, especially during the 1914-18 War. Only the meat was used, offal, heads and bones being simply thrown away. When slaughtering ceased abruptly after the war, the hyenas became really desperate. They bit the bristles off brooms, carried away pots, chewed and devoured any article of leather including shoes, sweat-stained hat-bands and bicycle saddles, rummaged through trash-cans, and ate a number of women working in the fields.

Cowie once hung a quartered gnu carcase high in the branches of a tree so that hyenas could not reach it, intending to feed the meat to some lions next day. He could hardly believe his eyes when one hyena, after leaping eight feet in the air, seized a gnu haunch and remained hanging there. A second hyena jumped up and sank its teeth into the leg of the first. The two animals swung there until eventually the rope broke and both hyenas fell to the ground with their prize. In other words, the teeth of the first hyena had supported not only its own weight but that of the heavy companion whose teeth had been fastened in its leg. This was the signal for other members of the pack to make similar incredible leaps. Very soon, one piece of meat after the other came tumbling down until the whole carcase had been devoured.

A rucksack containing cups, Thermos flasks, eating utensils and other articles vanished from the tent camp for tourists at Mweya in Queen Elizabeth Park, Uganda. Theft was suspected until the rucksack was found at the entrance to a hyena's den. Not long ago, a hyena chewed up and swallowed twelve plaster eggs in a hen-house at Serengeti. On the other hand, many misdemeanours are unjustly attributed to hyenas. A young hyena once disappeared from the Cros de Cagne zoo on the Riviera. As soon as this became known the proprietor had to compensate the owners of scores of dead rabbits, chickens, doves, and other domestic livestock every day – a sacrifice which he willingly undertook rather than get into even worse trouble with the police. After a few days the young hyena was found dead behind some packing-cases inside the zoo itself. It had never escaped at all.

Spotted hyenas make agreeable zoo inmates, even if few visitors find them particularly appealing. Their powerful teeth can crack marrow-bones which would defy the sharp fangs of lions and tigers, so their dung tends to be hard and easily becomes fossilized. Thanks to the latter and to many fossil bones, we know that Europe, too, was once inhabited by the cave hyena. This was bigger than the spotted hyena, the largest of the three species surviving today, but probably resembled it closely in many other respects. Zoo hyenas are long-lived. One specimen survived at the Berlin

Zoo for forty years, which would indicate that the hyena's lifespan is three times as great as that of the lion, tiger or leopard.

Most people are prone to compare hyenas with dogs, which are related to them. That is why their heads seem too clumsy, their shoulders too powerful, their backs too sloping, their hind quarters undersized, and their whole physical conformation ugly and distorted. Every species of animal has its own individual shape and beauty, however, and should not be judged by aesthetic criteria applicable to other species or human beings.

What is more, hyenas are far less like dogs than people originally believed. The spotted hyena has only two teats, and its young are correspondingly limited in number to two or, very rarely, three – never more. They are born after a gestation period of 99-110 days. In contrast to those of dogs and almost all other terrestrial predators, the young of the spotted hyena come into the world with their eyes open and are able to walk at once. They also have most of their incisors and all four canines, the latter being a good fifth of an inch long. Spotted hyenas are coal-black when born and do not become paler – starting at the head – until six weeks have passed. At nine months they are flecked like their parents. They become extremely affectionate, playful and amusing if trouble is taken with them. The Kruuks reared one in their house and let it roam free. Because it often ran into the hotel at Seronera, the headquarters of Serengeti, and frightened guests staying there, its foster-parents eventually locked it out in the hope that it would return to the wild and become independent. Undeterred, the powerful beast one day opened the front door by itself and joined Dr Kruuk's wife, who happened to be taking a bath, in the tub. Tame hyenas like to be stroked, scratched and fondled like dogs.

Dogs and wolves bay at the moon with their heads slanting upwards, hyenas with their muzzles pointing at the ground. One problem in zoos is that one never knows for certain whether hyenas are male or female. Superficially, their sexual organs appear identical. Many old settlers in Africa firmly believe in the myth

Hyenas can outrun zebras

that a hyena can alternately father young as a male and bear them as a female. According to Dr Wikingen it is possible to detect the penis-bone in a male hyena, but the only means of doing this – unless the animal is extremely tame – is to drug or X-ray it.

Spotted hyenas have frequently been trained for circus work, though one of them bit clean through the feet of its trainer, Trubka. Few travellers take hyenas very seriously, whereas sensations of fear are easily instilled by lions, leopards or rhinos. Their confidence is justified, on the face of it, because hyenas – like lions and leopards – will attack human beings perhaps once in ten thousand times. If tourists could once watch hyenas hunting and bringing down large zebras and gnus on a bright moonlit night, they would no doubt take a quite different view of them.

Spotted hyenas hunt either singly or in pairs and threes, but may also operate in packs up to a hundred strong. Individual hunters are less successful. Hyenas caught their prey in only four of the twenty-one single-handed chases observed by Hans Kruuk.

Page 137:
The spotted hyena is capable of speeds of 40 m.p.h. when pursuing prey and will cover 50 miles in a single day. Gazelles are the commonest prey of the 420 hyenas living in the Ngorongoro Crater.

Page 138, above:
Picturesque Fort Namutoni was completed by German colonial troops in 1904. Seven German soldiers successfully beat off an attack there by more than five hundred Ovambo rebels on January 28th of the same year. The fort was destroyed after they had withdrawn under cover of darkness. Three years later it was rebuilt in its present enlarged form. In 1957 the government of South-West Africa entirely restored this historic building. It is now a favourite overnight stop for parties of tourists visiting the Etosha National Park. In the foreground, makalani palms.

Page 138, centre and below:
This rare whitish-coloured giraffe lives partly in Tsavo Park and partly in the Mkomazi Game Reserve in Northern Tanzania. It has been repeatedly sighted for more than twelve years. Native game-poachers associate its colouring with the snow-capped peak of nearby Kilimanjaro and refrain from hunting the animal for superstitious reasons. Below can be seen a Burchell's zebra, also with unusual whitish markings.

Page 139:
This lion is showing great interest in the loudspeaker attached to the branch overhead, which is broadcasting the excited noises uttered by spotted hyenas dismembering their prey. We were able to lure lions to the most unlikely places by this means.

Zebras fight back

One solitary hyena finished off a full-grown gnu by drowning it in a lake. Of eight pack-hunts observed by Kruuk, eight achieved their object. Individuals generally bring down young animals and gazelles, whereas several animals join forces in order to hunt zebras or gnus. Of 188 hyena-dung samples tested at Ngorongoro, 83% contained the hair of gnus alone, 46% zebra-hair, and 16% that of Thomson's gazelles. 214 samples taken in the Serengeti plains contained gnu-hair in 54% of cases, zebra-hair in 30%, and Thomson's gazelle-hair in 53%. The reason for this may be that hyenas' tastes vary between groups. On the other hand, Thomson's gazelles are far commoner at Serengeti than in the Ngorongoro Crater.

Groups of zebras and gnus can attain a speed of 25 m.p.h. when trying to escape during night attacks, whereas hyenas are capable of 40 m.p.h. Moreover, most of the animals torn to pieces by hyenas are in good health. This is not to say that the spotted hyena and other predators do not prefer to destroy weak or ailing animals. It is merely that a wild herd does not contain enough weaklings to satisfy the requirements of all predators. Considerable numbers of young animals are taken. If they were not, grazing animals would soon become too numerous and ravage the country-side, creating a need for large-scale shooting drives. This, in turn, would make them as nervous of man as deer, hare, stags, rabbits and foxes are in Europe. Tourists would see them only in the distance or at night, if at all.

Page 140:
Wild animals seldom climb to the top of hills or mountains, hence this notice at a vantage-point in the Amboseli Game Reserve in Kenya. Using the tape-recorded cries of hyenas, we persuaded a whole pride of lions to congregate round the notice-board.

53 gnus were drowned while crossing the river at Seronera during mass-migrations through Serengeti in 1966. Accidents of this type can occur when panic suddenly breaks out in the closely packed herds. A pride of seven lions recently prevented gnus from drinking in the Kruger National Park. One female gnu was caught between the lions and the water. She galloped wildly through them and raced between visitors' cars with the predators close at her heels. Having attained the safe distance of two hundred yards, she turned and eyed her pursuers, who had flopped down and were gasping for breath.

Human beings killed by hyenas

Hunting hyenas generally seize their prey by the leg or flank, sinking their teeth in and hanging on until the animal falls to the ground. They at once tear off the soft flesh between the hind legs and on the belly. The prey-animal survives for a while and tries to defend itself. When more than ten hyenas tear at a zebra or gnu, the animal usually sinks to its knees after four or five minutes and is dead within about ten. On the other hand, I have often watched one or two hyenas hunting, catching and dismembering small gnus or gazelles in the midst of a scattered herd. The mother or other full-grown animal may try to close with the predators and drive them off, but the rest of the herd takes little notice. Where zebras are concerned, the stallion of a family will almost invariably defend its young with bared teeth and flailing hoofs. In general, the mares of the family group assist in trying to repel an intruder. The game warden Myles Turner once saw a hyena pursue a common jackal into the middle of a herd of Thomson's gazelles and catch it after four minutes. Young gnus born at the beginning of the rainy season have little prospect of survival. Only when all female gnus produce young almost simultaneously are their numbers such as to defy extermination by hyenas and other carnivores. In the course of one very dark night at Serengeti hyenas wrought real carnage among Thomson's gazelles. They killed more than a hundred at one spot but devoured only a handful of them.

In places where wild grazing animals have been virtually exterminated by poachers or land is being cleared for future habitation, hyenas can become extremely dangerous to man. In the Mlanje district of Malawi (formerly Nyasaland) three people were killed by spotted hyenas in 1955. The first was a child of whom only the head was left, the second a woman who was seized by a hyena and dragged for ten feet. She screamed for help and was rescued by fellow-villagers, but had lost an arm and was so bitten about the throat that she died a few hours later. Her attacker was probably the hyena which had killed the child, because the two incidents occurred within seven miles of one another. The third victim, a mental deficient, was killed on the way from one village to another.

He was devoured, once again by a single hyena which weighed 156 lb. when shot. Five inhabitants of the same district were killed by hyenas in 1956, another five in 1957, six in 1958, and no less than eight in 1959, most of them from September onwards. The inhabitants of the area sleep inside their huts in winter, not outside, which is why there are no casualties in the cold season. Spotted hyenas will not as a rule touch hyena-meat, though two instances of cannibalism have been reported by F. Balestra. Mervyn Cowie, who forty years ago had to exterminate more than a thousand hyenas in one district, told me that the animals do sometimes bite each other to death when fighting over females or prey. On the other hand, a spotted hyena never touches the carcase of one of its own species until decay is far advanced, which takes at least four days. Although vultures begin their meal immediately, other hyenas do not join in.

One spotted hyena seized a woman on a farm near Mount Darwin in Rhodesia. Although help arrived the woman was dead within a few hours. The same day a man was accompanying his child home from school. On the way, the child was attacked by a big hyena. Its father managed to beat off the animal with an axe but was bitten in the hand. Another man who ran to help was bitten in the leg. The hyena received a heavy blow on the head with the axe and was later shot. A few years ago, a German schoolmaster who had failed to find accommodation in the tourist hotel beside the Ngorongoro Crater was sleeping in the open not far away. He awoke in the middle of the night to find himself seized by the leg and dragged along the ground. His first thought was that a lion had got him. Some people in the vicinity answered his cries for help and put the predator to flight. It turned out to be a large hyena, which had grabbed his leg through the padded sleeping-bag and dragged him away. The leg was badly mauled.

Hyenas can come to grief when hunting in packs. Seven spotted hyenas chased a male oryx antelope for three miles through the Kalahari National Park in South Africa. The oryx killed one of them with its long pointed horns but was finally dispatched by the other six. Myles Turner saw a male lion kill a big hyena at Serengeti and then stalk two others in quick succession. Lions do not take much notice of hyenas as a rule. On another occasion, however, hyenas in Kruger Park chased two young lions up a tree and did the same to a full-grown lioness. Turner also saw a hyena sleeping beneath a tree immediately beside a pair of slumbering wart-hogs, so close that it was actually touching one of them. A lone elephant, trumpeting with rage, was seen to pursue a hyena along a sandy river-bank until the hyena succeeded in scaling the bank and escaping into the bush. The elephant's motives are not known. Cherry Kearton watched a hyena and a python fighting near a papyrus swamp. The big snake was unable to exert its full strength and get a firm grip on the hyena with its teeth because there was no suitable purchase in the vicinity. It was so badly lacerated by its opponent that it eventually collapsed and died. At Serengeti, when four of the redoubtable African wild dogs had caught a young hyena, twelve big spotted hyenas ran up and succeeded in rescuing their small companion by driving the dogs away. The African game warden N. L. Ochara told me that people in his native Uganda believe hyena's gall to be so poisonous that it causes instant death. The meat is thought to cure framboesia, a tropical disease of the skin. Possession of a dead hyena's nose imparts an excellent sense of smell and enables blind men to find their way without a guide. On the other hand, nobody carrying such a nose must ever visit the sick or they will die. Meal ground from the first rib of a hyena brings luck in business and also heals chest complaints. Before we snigger at such African superstitions, let us remember the astrological advice purveyed by European magazines and newspapers!

Apart from the big spotted hyena, which lives in Africa south of the Sahara, there are two more largely nocturnal species: the brown hyena of South Africa, and the striped hyena, which is found in North Africa and South Asia. Smaller and more timid than the spotted hyena, striped hyenas were tamed and used for hunting by the ancient Egyptians, who even fattened them for slaughter. One cave examined by C. K. Brain and used for defecation by brown hyenas contained ninety identifiable bones belonging to prey-animals. Sixty of them were hares and dassies. The spotted hyena is the only species whose living habits have been studied at all closely, and that only in recent years. Our information about the other two species is as scanty as it is in the case of so many other wild animals.

9. Jumping giraffes

Dying men are kind to each other; dying cows lick each other.
AFRICAN PROVERB

How would you transport an obelisk weighing many tons – horizontally on rollers or balanced carefully on end to prevent the tall column from toppling? Nature has generally favoured the first alternative. Long flat creatures like snakes and lizards propel themselves along without effort. Travelling vertically erect becomes more difficult the taller one is. The giraffe is the tallest of all living animals, as much as 20 feet in height, anything up to 2,600 lb. in weight and extremely short-winded besides. Its heart has to pump blood seven feet in order to reach the head, which is why the walls of its carotid artery are 1.2 cm. thick and its blood pressure is twice as high as that of a human being. The Leningrad Physiological Institute has established that a giraffe's lung will hold only 12 litres of air compared with the horse's 30. Consequently, giraffes on the run can easily be overhauled by a man on horseback and chased until they come to a standstill or collapse.

American scientists fitted the carotid arteries of two large giraffes with a gauge coupled to a self-contained transistorized radio transmitter. A receiving set installed in a field-car evaluated information as it came in. Findings seemed to indicate that the heart-beat and blood pressure of a giraffe are subject to exceptionally wide fluctuations. The heartbeat between 90 and 100 times a minute in repose and 170 times when the animal was travelling at speed. What was more, its blood pressure fluctuated rhythmically during gallops every time the forehoofs touched ground. It mani-

Browsing can be dangerous

festly rose and fell as the rise and fall of the animal's head affected the pressure in its carotid artery.

Goats and antelopes balance on their hind legs in order to reach high branches. A giraffe remains permanently on all fours because it could never perform such a feat. How, one wonders, could a creature which keeps its feet planted so firmly on the ground have managed to get the better of wire fences? Giraffes in the Transvaal began by marching straight through with broken wires trailing behind them. Then, after three or four years' experience of the new things, they astonished local farmers by learning to jump them. In the end, even a 6-foot barbed wire fence held no terrors for them. Throwing back their heads and necks, they cleared the obstacle with their forefeet and did not worry unduly if their hind legs caught the top few strands of wire. Fences occasionally proved too much for calves. One female which had jumped a fence with the herd found that its offspring was marooned on the other side. The mother paced up and down the barrier for a full sixteen hours without taking it into her head to jump back on her own. The calf finally disappeared into the bush, probably for good.

Giraffes have even learnt to drink from concrete troughs intended for herds of domestic cattle, something which many other wild animals cannot bring themselves to do.

In January 1963 the administrators of the Kenya game reserves overcame considerable difficulties in order to transport four giraffes from the Eldoret district to the former Menengai Crater, 100 miles from the capital, Nairobi. One of them, a bull, was about to browse on the upper branches of a tree in its new home when it disturbed a leopard which happened to be sleeping there. The leopard promptly sprang at the giraffe and mauled its neck so severely that it died.

In general, giraffes live at peace with their environment. Just occasionally, though, peculiar ideas seem to enter their lofty heads. A doctor driving with his family at dawn near Rusermi in South Africa met two giraffes coming in the opposite direction. He stopped his car and switched off the headlamps for fear of dazzling them. One of the animals left the road, but the other strode calmly and resolutely up to the car, which was still ticking over. It then turned round and started to belabour the vehicle with its hind legs. The radiator and front of the car sustained at least three powerful and well-aimed kicks before the stupefied doctor stepped on the gas and drove to safety. He was so proud that his car had been damaged in this unique way that he drove it around for quite a while before having it repaired.

In 1963 a field-car belonging to the geological research department was attacked by a giraffe while returning after dark to Moroto, capital of the Karamoja district of North-East Uganda. The driver pulled up as soon as he turned a corner and saw the animal standing in the glare of his headlamps some twenty-five yards away. The giraffe turned and struck at the vehicle with its forelegs, smashing the headlamps and windscreen and bending the steering-wheel. It eventually leapt over the car and disappeared without injuring any of the passengers.

I myself have encountered giraffes a dozen times while driving cross-country or along roads, but none of them evinced the least intention of picking a quarrel with my car. Both these encounters would therefore seem to be instances of quite exceptional behaviour.

The skyscrapers of the animal world are even less well equipped to cope with mountainous country, which is why they tend to avoid it. Not long ago, at the Ngurdoto Crater in Arusha National Park, one of the new game reserves created by the independent government of Tanzania, an old bull giraffe was seen ascending the lip of the splendidly luxuriant inner crater by way of a precipitous track used by Cape buffalo. Because heavy rains had just fallen and the path became steeper and steeper, the animal began to slip with increasing frequency and finally fell flat on its face. Having laboriously regained its feet, the giraffe found itself confronted by a troop of baboons. They glared at it resentfully because it was barring their downward route to the water-hole in the crater. Nothing happened

A giraffe killed by a lion

for quite a while. Then one half-grown baboon dodged impudently between the giraffe's legs. The giraffe lashed out but failed to connect and landed flat on its belly again. As soon as the other baboons saw this they poured down the path like a waterfall, brushing past or underneath the giraffe, which had struggled to its feet and was standing there motionless. As soon as all the baboons were through, it turned and 'sadly' slithered back the way it had come.

Giraffes are seldom encountered singly. Closer scrutiny will usually reveal other giraffes browsing among trees and bushes in the vicinity. A lone giraffe is more vulnerable to attack by lions, which generally show little inclination to tackle prey of such magnitude and fighting spirit. The game warden De La Bat once watched a lion stalking a lone giraffe in the Etosha Pan, a district in South-West Africa. The giraffe made off as soon as it noticed its enemy, but the lion overhauled it after a few strides. It leapt for the neck, dug its claws in, and probably severed the giraffe's cervical spine with its teeth. The big animal staggered and collapsed. On another occasion, the attacking lion failed to get close enough before the giraffe noticed it, and had to pound along for some distance before overhauling its prospective prey. Lions have little staying power, so this one was probably rather exhausted by the time it launched itself at the bull giraffe's back. It landed on the rump, slid off, and caught two flailing hind legs full in the flank. A game warden shot the injured predator some hours later, when it showed no signs of recovering. It turned out that the lion's chest had been crushed and almost all its ribs broken.

Giraffes never lash out with their legs when fighting among themselves. Not long ago I spent several hours watching ponderous battles between bull giraffes at Serengeti. It is evident that giraffes, too, have a ranking order, but they usually live together in such harmony that a master bull and its inferiors are barely distinguishable. Perhaps the best indication of seniority is that the senior animal will balk a junior by simply cutting across its line of advance. Master bulls also carry their heads higher and their chins more tilted. An inferior inclines its head slightly when a superior

Giraffe's horns, short but effective

passes by. One giraffe impresses and threatens another – or, in a zoo, its keeper – by raising its head.

No young bull giraffe is permitted to 'fleer' in the presence of a superior. Fleering is an expressive movement – commonly found among horses, camels, bears and many other animals – in which the head is slightly raised, the jaws slightly opened, and the upper lip drawn back. It probably has some sexual connotation. A giraffe which seriously intends to drive another away walks straight up to it and lowers its head threateningly. The giraffe's short horns are its main weapon in contests between two bulls, together with its large and heavy head, which can deal terrible hammer-blows. The horns are encased in skin and hair except in the case of bulls, whose blunt horn-tips are abraded and bare of hair. This is one way of identifying the sexes.

It is quite possible to live in Africa for decades, even in the bush, without ever seeing two bull giraffes engaged in serious combat. This is some indication of the rarity with which intraspecific fighting occurs, certainly in many areas. It is an exciting spectacle, even though everything seems to happen in slow motion.

In contrast to stags, bull antelopes and feline predators, which fight facing one another, giraffes usually stand side by side, either with their heads pointing the same way, like a team of horses, or nose to tail with their heads pointing in opposite directions. They never try to bite one another. Instead, they draw back their necks in a wide arc and smash their skulls against an opponent's head, neck, chest or exposed flank, or, when standing nose to tail, against rump and hind legs. The force of such blows is almost unimaginable. Otto, our bull giraffe at Frankfurt Zoo, swung at a big bull eland which had been pestering him. The eland, a massive animal weighing more than half a ton, was propelled through the air with such force that it broke a shoulder and had to be destroyed. A cantankerous bull giraffe at London Zoo once lashed out at its keeper and missed, knocking a hole in the wall behind him.

Ranking-order duels can go on for half an hour or more. Other bulls and females are often present, but fighting also takes place in

Victories not pressed home

the absence of female onlookers. At Seronera in Serengeti I was able to film three pairs of bulls fighting simultaneously. Although only 80 or 100 yards apart, they took no notice of each other. The rest of the herd continued to browse all round them. A fighting bull sometimes forces its opponent against a tree, or the two animals may circle a tree-trunk as they fight. As soon as one of them recognizes its inferiority it withdraws a few paces. The victor follows with head raised, but not for very far. Stags, antelopes and many feline predators drive vanquished males from their territory and do not allow them to return, whereas bull giraffes generally resume friendly relations once their trial of strength is concluded. They have even been seen to rub necks immediately after a fight or resume browsing peacefully together.

Fights of this kind seldom end in serious injury, let alone death. Fatalities are probably unintentional accidents of the sort that occur in human boxing matches. The swellings and protuberances quite commonly found on giraffes' necks may well be caused by fighting. I myself have come across a number of such animals in the northern part of Serengeti National Park. In 1958, a bull giraffe sighted at Kruger National Park in South Africa was reported to have sustained 'a broken neck'. The injury, which had healed, was located about $2\frac{1}{2}$ feet below the head. The animal was in good health but the neck 'looked as if it might break again very easily' – a somewhat bold assertion in view of the impracticability of closer examination. On another occasion, also in Kruger Park, a bull giraffe was found lying dead 'as a result of a mamba-bite'. This is a favourite explanation for the discovery of dead animals which are in good physical condition and show no sign of injury. Personally, I find it hard to believe. There is no doubt that a large beast weighing more than a ton can ultimately die of mamba-venom, which very swiftly causes death among the smaller animals that are the poisonous snake's natural prey. Even human beings generally take hours to die, if not days. Their bodies are not unmarked, either. Snake-bite produces severe swelling in the affected limb, bleeding from all external bodily orifices, tissue decay, exposure of

The dead giraffe that came to life

bone, and haemorrhagic discoloration of the skin. In the case of an animal so large that the venom is far more diluted, one would expect the latter to cause far greater ravages before death intervenes. Fractured skulls and broken necks caused by well-aimed blows inflicted during intraspecific fights are therefore a more likely explanation for the death of the very few giraffes found without external injuries. In 1958 a Mr Innes saw a bull giraffe knocked out for twenty minutes in the course of a fight. The base of the giraffe's neck, with its large blood-vessels, is particularly vulnerable. In giraffe-hunting, which is now universally banned, hunters always used to aim for that part of the body.

In July 1958, near Heilbron in South Africa, a party of tourists found a full-grown giraffe stretched out on the ground about thirty yards from the road. Believing it to be dead, they got out to inspect it, only to discover that the animal was breathing and looking at them with its eyes open. When it still failed to move, the male members of the party took their courage in both hands and rolled the animal on to its stomach, presumably in ignorance of the dangers of such a procedure. Zoo staff would have been considerably more cautious. The giraffe, which offered no resistance whatsoever, eventually stood up of its own accord. Once on its feet again, it looked down 'haughtily' at the amateur photographers and camera-men that surrounded it. Then it switched its tail and strode calmly off into the bush, casting an occasional backward glance at its helpers.

At Serengeti in April 1963 we came across the unmarked carcase of a female giraffe which had died while giving birth. A lone female at Kruger National Park gave birth well within sight of the road. She was standing in a small dip beneath some trees, with the neck and forelegs of the calf already visible. The expectant mother remained quite calm, taking no notice of some onlookers who had stopped their cars only twenty yards away. From time to time she suffered contractions and forced the calf out, inch by inch. Finally, the calf's shoulders appeared. After further pressure they slid out followed immediately by the rest of its body. The umbilical cord snapped as the calf fell to the ground. The mother soon splayed her legs and bent down to inspect the new arrival, which endeavoured to stand up after about fifteen minutes. It continued to struggle in vain while its mother browsed contentedly on the surrounding trees, glancing down at it from time to time. During another giraffe birth at Kruger, no less than nine giraffes formed a circle round the expectant mother. On this occasion the calf succeeded in standing after about ten minutes but fell over again. These exertions were repeated over a period of some twenty-five minutes, after which time the young giraffe walked round its mother and seemed quite strong. All the other giraffes bent to nuzzle the new addition to the herd.

Young male giraffes often fight for fun. Otto, our bull giraffe at Frankfurt Zoo, liked to indulge in lengthy mock-battles with his son Thulo. In default of a real opponent, Otto also used to butt the suspended food-basket with his head. On another occasion, tourists reported seeing a fight between two giraffes on the Malelane road in Kruger Park. Hearing that one of the animals had been killed, Game Warden de Clerck drove to the scene at once and found a big bull giraffe lying dead beside the road. It had a huge hole just behind one ear and had lost a good deal of blood. The topmost vertebra was smashed and fragments of splintered bone had pierced the animal's spinal cord.

It is hard to imagine what goes on inside the giraffe's bony head or how the world looks to eyes that gaze down on it from the height of a first-floor window. According to exhaustive research undertaken at Frankfurt by Dr Backhaus, there is no doubt that giraffes see things in colour. They can certainly distinguish between the basic colours, yellow, blue, green and red. Their inclinations seem to be little affected by personal appearance. At all events, the same herd often includes giraffes which vary widely in marking and coloration. At Serengeti I repeatedly, and over a longish period of time, sighted a very dark bull giraffe whose coloration verged on black. White giraffes have been observed in the Garamba and Murchison National Parks, and sightings were made in 1963 of

White giraffes

another which was browsing peacefully among giraffes of normal coloration in the Rukwa Game Reserve in Zambia. The latter animal was actually photographed from the air. In 1965 a white giraffe gained notoriety in the Mkomazi Game Reserve in Northern Tanzania, which adjoins Tsavo National Park in Kenya. The animal in question is an old bull which keeps to itself and is never seen in the company of other members of its species. Its eyes are dark, in addition to some other features, so it cannot be a pure albino. Its existence has been known of for over ten years, because it used earlier to be seen in Tsavo Park, some eighteen miles from its present place of residence. Fortunately, this rare specimen seems to be safe from the attentions of native poachers. They refrain from hunting it for superstitious reasons because they associate its colouring with the spirit-infested peak of near-by Kilimanjaro. An entirely unmarked giraffe was born at Tokyo Zoo in 1967.

10. Crocodiles that can remember Stanley and Livingstone

If you wade in the lagoon, abuse not the mouth of the crocodile.

AFRICAN PROVERB

More and more people travel to Africa each year in order to see and photograph the large and impressive wild animals of that continent, comfortably but at close range. As I have already said, there is only one place where they can be reasonably sure of getting a good view of large crocodiles, and that is Murchison Falls National Park, where, snugly seated in a motor-boat, they can photograph them basking on the sand-banks of the Victoria Nile below the Falls themselves. One can spend days surrounded by wild life in Uganda's second large national park, the Queen Elizabeth, without seeing a single one. This is because the Nile crocodiles and a number of other aquatic species widely represented elsewhere were destroyed by volcanic eruptions which convulsed Lake Edward in prehistoric times. The crocodiles have never succeeded in returning to Lake Edward from Lake Albert via the Semliki Nile, which links them. Although this may be attributable to the waterfalls in the middle of the jungle, it is more likely that the big saurians find the water fed into the Semliki by tributaries carrying melted snow from the Ruwenzori Mountains disagreeably cool. However, big crocodiles have become rarer in the Victoria Nile as well, and the concentrations of earlier decades no longer exist.

Crocodiles are interesting creatures, and there is a definite need to ensure their survival. They used to be found in every river and lake in Africa. The last crocodiles in Palestine, modern Israel, were not exterminated until the first decade of the present century. A

The threat of extinction

few stunted and lonely specimens still inhabit the waters of some oases in the Southern Sahara. Crocodiles have always been hunted because they devour domestic animals and the occasional human being, but they were never in danger of extinction for that reason. What has really dealt them the coup de grâce, apart from modern fire-arms, is the recent vogue for crocodile handbags and shoes, which has boosted the price of skins to record heights.

It has become profitable to comb the most remote rivers and marshes by night, dazzling the creatures with beams of light and firing at the reflected glow of their eyes. Even tiny little crocodiles far short of sexual maturity bring good money as tourist souvenirs when stuffed. One white hunter was until recently making an annual income of £25,000 by hunting crocodile in the Okavango Marsh in Botswana. He bought his permit from the Batawan tribe and the government for fees of £3,500 and £1,200 respectively. In New Guinea, also in recent years, the very species of crocodile most in danger of extinction was assiduously hunted by a forty-one-year-old nun who planned to use the proceeds to build a small mission church. Game wardens in Florida are waging a desperate and hitherto fruitless campaign in protected areas to protect surviving stocks of alligators against organized gangs of poachers.

Page 157:
Giraffes lash out at lions and other predators with their hoofs. They are quite capable of killing a lion. Among themselves, bull giraffes deal each other powerful blows with their hard, heavy skulls and short horns. They fight to prove their superiority but do not, in contrast to stags, male antelopes and many other species, pursue a worsted opponent. It is very seldom that one can observe, let alone photograph, one of these fierce duels between bull giraffes. Bulls may be as much as 20 feet tall and weigh over a ton. A hard, well-aimed blow can knock a bull senseless for up to twenty minutes, but its opponent at once leaves it in peace. The other members of the herd do not interfere, merely watch the fight or continue to browse.

Pages 158-9:
I quite unnecessarily broke a rib while photographing this herd of giraffes. I was standing on the platform of a field-car as it drove parallel with the galloping herd in the Kirawira Plain outside Serengeti National Park. One of the front wheels went into a warthog-hole, hurling me against the steel framework which normally supported the canvas roof. The steel bar bent quite considerably, but I was not badly hurt. This area has recently been added to the Serengeti National Park by Dr Julius Nyerere's new government. A number of African villages had to be cleared for the purpose.

The dangers of research

Being unafraid of salt water, crocodiles even colonized the island of Madagascar, far out in the Indian Ocean. The bones of extinct Nile crocodiles thirty feet long have been found there. Crocodiles are extinct in the Nile itself as far south as the Second Cataract. They were so numerous at the beginning of this century in German East Africa, now Tanzania, that the government paid a premium of $\frac{1}{4}$-3 rupees per head. In 1910 a cattle dealer set off for Lake Rukwa and earned himself 5,000 rupees in premiums. As late as 1950, 12,509 crocodile-skins were taken in what is now Tanzania, mainly in Lakes Rukwa and Victoria and in the Ruvu river. Yet the crocodile is an important link in the chain of Africa's aquatic wild life, and the long-term effects of its disappearance on the continent's stocks of fish have still to become apparent.

Hugh B. Cott of Cambridge University has devoted five years of his life to studying crocodiles in Zambia and Uganda. In 1965, the Indo-African biologist M. L. Modha spent eight months on Central Island in Lake Rudolf, Northern Kenya (see p. 190 et sqi.) in night temperatures of 26°C. (78.5F.) and day temperatures ranging between 36°C. (96.5F.) and 42°C. (107.5F.) While engaged in further research during 1968 he was seized by a smallish crocodile and badly mauled. An African companion was fortunately able to release him. Central Island measures $3\frac{1}{2} \times 2\frac{1}{2}$ miles, and one of its three volcanic lakes is inhabited by about 500 large crocodiles. Being between

---

Page 160:
The ground squirrel usually carries its long bushy tail low when it runs. Having reached the entrance to its home, however, it often rises on its hind legs to take a last look at an intruder before vanishing into the burrow, which may be as much as five feet deep. These little creatures live on a vegetable diet and their burrows are interconnected. When Cape buffalo fight, the horny protuberances on their foreheads clash together with a sound which carries a long way. The animals shove and butt each other back and forth with their massive helmets of horn. The really dangerous points are not used in these intraspecific trials of strength, so blood is seldom shed. On the other hand, the curved ends of the horns can interlock and snap off. Arusha Hospital in Tanzania once admitted a Masai tribesman whose chest had been pierced by a spear. Masai spears are equipped not only with a three-foot metal point but also with a short sharp metal butt which serves to hold the weapon upright when driven into the ground. The Masai in question had attacked a Cape buffalo with his spear, but the animal continued to bear down on him with the point in its body and drove the butt-end through his lung. Dr Eckhard, the medical superintendant, informed me that the Masai recovered.

Territorial defence

11½ and 13 feet long, the males are readily distinguishable from the females. They have divided the 1,300 yards of lake-shore under observation into 12 territories, and will chase intruders to 50 yards off-shore. The females favour males with the best places for basking and nesting, not those that are largest.

Territory-owners swim the length of their stretch of shore even during the hottest period of the day, when all other crocodiles are basking and resting on land. They always make a five- or ten-minute halt at the same spot on the borders of their territory before turning and swimming back again. Territorial borders may be displaced by male intruders or as a result of fighting. Unlike stags, for instance, male territory-owners do not herd their females together but simply drive off other males of the same species. Trespassers flee for the shore with their snouts projecting above water, pursued by the owner of the territory in question. The latter only half emerges from the water in order to roar and snap in their direction. A few minutes later it backs off and continues to swim up and down its strip of shore.

In general the big lizards spend their nights in the water and most of the day sunning themselves on shore. At midday they usually seek out some shade or cool off by taking a short dip, though Central Island is so hot that this period is extended to a full three hours. There is no sign of them on shore by nightfall. Although referred to by text-books as cold-blooded or at least variable in temperature, crocodiles do by this means succeed in maintaining a fairly even body-temperature of 25.6°C. (78F.) which fluctuates at most by 3.4°C. upwards and 2.6°C. downwards. They often open their mouths wide during the midday heat. Having no sweat-glands, they thus evaporate moisture from their oral membranes and cool themselves as we and other mammals do by perspiring. Lake-dwelling crocodiles are seldom if ever found in open water. Even when Lake Victoria was still well-stocked with them, one had only to take a boat a couple of miles out to be able to bathe without fear.

The crocodile's gaping jaws prompted Herodotus (490-420

B.C.) to tell how a small bird called the trochilus used to slip into its mouth and peck off the leeches there. This story was repeated by numerous ancient writers. Pliny claimed that the bird also extracted pieces of meat lodged between the crocodile's teeth because it had no tongue with which to clean them itself. This last assertion is certainly inaccurate. Nile crocodiles do have tongues, but they are broader than ours and cannot be extended. Aristotle went so far as to state that crocodiles moved their heads before shutting their mouths to warn the birds and give them a chance to escape. Aelian added that the bird's cry warned its crocodile host of impending danger.

This is quite true. The spur-winged plover (Hipoplopterus spinosus) has only to utter its shrill warning cry and all crocodiles in the vicinity at once slide into the water. This bird is also privileged to hop about on and among crocodiles with impunity. The common sandpiper (Actitis hypoleucos), which breeds in Europe and visits Africa in winter, collects the parasites on crocodiles' bodies and positively scampers to meet them as they emerge from the water.

Modern zoological works do not, however, accept the statement that birds enter the crocodile's open mouth and clean its teeth. I have never witnessed such a procedure myself, though J. J. Player describes how he actually watched a sandpiper gathering food in a crocodile's mouth in Zululand. It apparently stood on the lower jaw and picked leeches off the creature's oral membranes. The Swiss biologist Guggisberg, who also doubted the story, saw a marabou thrust its beak three times into the open jaws of a crocodile on the Victoria Nile and finally bring out a small fish. This incident was actually filmed.

The water dikkop (Burhinus vermiculatus) always broods a few yards from the spot where crocodiles have buried their eggs and are guarding them. The crocodiles ignore the bird, which benefits from their tutelary presence.

Crocodiles have three forms of locomotion. Although they can slide on their bellies with legs splayed sideways so as to reach the

The mystery of the pebbles

water quickly, they usually proceed on land at a waddle, turning their legs in slightly so as to raise their bodies off the ground. They can also gallop, at least when young, although they do so very rarely. This gait resembles the hopping of a squirrel. In water, crocodiles swim like fish, clamping their limbs to their sides and propelling themselves along by wriggling and beating their tails. Small crocodiles less than 3 feet long can remain on the bottom for up to 44 minutes, full-grown specimens for over an hour.

Many theories have been advanced to account for the stones found in crocodiles' stomachs. It seems unlikely that they serve to break down food, as in the case of many birds, because crocodiles have no gizzard with walls capable of trituration. Besides, stones are found adjacent to undamaged round-worms, mussel-shells and similar objects. They are never present in young crocodiles in the first year of life but almost always in adults, being so numerous that they account for about 1 per cent of total body-weight. When a large reptile of this kind enters the water, buoyancy reduces its weight by 92.5 per cent, so a creature weighing 400 lb. weighs only 30 lb. when water-borne. H. B. Cott, the authority on crocodiles, argues from this that the stones are a form of ballast.

Crocodiles are extremely tenacious of life. The big game hunter Alexander Blake, who wrapped the heart of a large male crocodile in a damp cloth and put it in the sun, found that it was still beating half an hour later. He once watched between four and five hundred crocodiles thread their way past his camp in a single night. The German colonial officer Hans Besser, who spent fourteen years in East Africa before World War One, one day shot a 7-foot crocodile which nevertheless escaped into the water. His courageous African bearer, Mtuma, plunged in after the reptile, seized it by the tail, and dragged it ashore. It continued to resist, sending Mtuma and Besser himself, complete with rifle, head over heels into the water. Once ashore again, Mtuma nailed it to the ground with an iron buffalo-spear. When the crocodile still opened its jaws, Mtuma withdrew the spear and thrust it between them as far as he could into the creature's body. The crocodile clamped its jaws shut and held the iron spear so tightly between its teeth that Mtuma had to tug it out inch by inch, the iron shaft grating horribly as he did so. He then cut into the crocodile's neck with a hunting-knife and transfixed the spinal column with his spear-point. 'Because my friend had not turned up,' Besser wrote, 'I scribbled a note informing him of my departure and stuck it in the crocodile's mouth. Then I dragged the crocodile across the path so that my friend would be bound to see it when he arrived. After I had washed my hands I sat down in the shade of a tamarind and had breakfast. Then Mtuma said: "There, sir, just look at your crocodile!" And indeed, I could hardly believe my eyes, but the crocodile made for the water in leisurely fashion and disappeared despite the severe injuries we had inflicted on it, each one of which should have been fatal.'

Full-grown crocodiles live mainly on mammals. R. J. G. Attwell saw them follow floating animal carcases for miles down the Luangwa river in Zambia (formerly Northern Rhodesia). They even devoured dead elephants in the water, though crocodiles in this area lived more on Cape buffalo, which often drowned and sometimes perished by the hundred during floods. Food was also provided by male hippos killed as a result of intraspecific fighting. The crocodiles were extremely nervous at first, retreating from a carcase when approached too closely by someone on shore or when an observation-hut was erected. As soon as the big vundu fish (Heterobranchus longifilis) or vultures set to work on the dead animal, however, greed overcame their fear and they quickly returned to the scene. On one occasion 120 crocodiles busied themselves with a single hippo carcase, so that the river was emptied of them for two miles downstream. On arrival, crocodiles swim round a carcase until their heads point upstream. One crocodile ventured twenty yards inland to fetch the entrails of a buffalo which had been killed by a lion.

Crocodiles do not, in fact, possess teeth which are suitable for tearing at or chewing large prey. In the case of a freshly-killed hippopotamus or buffalo, the hide is too tough for them to do more, at first, than bite off the ears and tail. They do not reject fresh meat of suitable shape but are incapable of biting off mouthfuls. This is why they often push dead antelopes into under-water holes or store them beneath overhanging river-banks until decay sets in and the hide begins to soften. J. Stevenson-Hamilton, a game warden at Kruger Park, reports that villagers probing a river with sticks found the almost unmarked body of a young African boy tucked in beneath an overhanging section of bank.

Crocodiles tear mouthfuls of meat off a carcase by sinking their teeth in it and jerking round abruptly in the water. The tail usually breaks surface, exposing the pale under-belly. This twisting motion may be repeated up to nineteen times in succession, often with such force that the crocodile turns completely on its own axis. Having ripped off a mouthful it rises so that its head is only half-submerged, as we were able to observe in the twilight at Serengeti, where a crocodile was devouring a half-grown zebra. The mouthful is then jerked farther down the throat. In the above instance the crocodile remained on the surface for several minutes after each mouthful, breathing heavily, before submerging to take another bite below water (Myles Turner). Between thirty and forty crocodiles were once sighted round a dead buffalo in the Luangwa river, but there was very little squabbling or biting. It is interesting to note that small and half-grown crocodiles 8 feet long or more are never seen together. Large fish which are normally preyed upon by crocodiles often jostle them for a share of the spoils. One extremely large fish was actually hurled out of the water when a crocodile swung round.

The crocodile's digestive system is not very efficient or fast-working, which may partly explain its predilection for decaying carcases. The *Times of Swaziland* reported on November 8th 1952 that a $13\frac{1}{2}$-foot crocodile had attacked an eight-year-old boy on the shores of an island in the Usutu river near St Philip's Mission, seizing him by the buttocks and dragging him into the water. A nineteen-year-old youth immediately crossed the river and stabbed the crocodile several times with his spear. Darkness soon fell, unfortunately, and it was next morning before he and the child's grandfather could track the creature down. They eventually found it lying on a sand-bank. The young man threw his spear from a distance of 20 feet, whereupon the crocodile took to the water with the shaft protruding from its shoulder. Its pursuers finally cornered it between branches and sticks on the other bank. The courageous young man waded into the water and continued to attack the crocodile until he killed it. When the creature was hauled ashore and cut open, the upper half of the child came to light in the stomach, quite recognizable and not in the least digested although it had been inside for twelve or fifteen hours. The boy's grandfather buried him and placed the crocodile's severed head above his grave. The jaws were propped open with a stick.

One demonstration of the crocodile's stength was given by a 13-foot specimen lying on a sand-bank in Natal. When disturbed by a full-grown female impala, it hurled the antelope (which weighed approximately 1 cwt.) to the other side of the bank with a single jerk of its head and vanished into the water like greased lightning.

During 1927, one farm in Southern Rhodesia lost 179 cattle to crocodiles in the course of ten months.

People have always been deeply impressed by the crocodile's ability to kill human beings. The ancient Egyptians believed that their river-god must be appeased by the annual sacrifice of a beautiful virgin. The unfortunate girl was thrown to the crocodiles and devoured in the course of a public festival. The inhabitants of the Sese Islands in Lake Victoria used to regard the crocodile as the high priest of a god. Sacrifice took the form of breaking people's arms and legs and leaving them on the shore where the crocodiles could help themselves. During religious wars, King Mutesa marooned African prisoners of the Muslim faith on small islands in Murchison Bay, where they either died of hunger and thirst or were inevitably torn apart by crocodiles while trying to escape.

My son and I visited the Dan, a West African people distributed between Liberia and the Ivory Coast. Dr Himmelheber had made friends with a local chief and was thus able, partly in Africa and partly while the man was visiting him in Heidelberg, to undertake a thorough study of the Dan religion. The Dan regard a crocodile's gall as poisonous. A hunter who has killed one must notify his village at once. The witch-doctor then extracts the gall-bladder intact, cuts it up in full view of the villagers, and allows the contents to flow away down-river. This means that nobody can use the supposedly poisonous gall for nefarious purposes.

In March 1959, at Chikwawa in Malawi (formerly Nyasaland), an old man prevailed upon someone called Elard Chipandale to kill his granddaughter. Chipandale, a notorious 'crocodile-man', was to do away with the eight-year-old child because she refused to obey her father. The murderer received one payment of £2 on account but took the grandfather to court when the remainder of his fee failed to materialize. He informed the astounded judges that he could transform himself into a crocodile by strapping magic branches to his hips, whereupon he acquired a long snout, a long tail and sharp teeth. It was the tail, he said, that had broken the child's bones. When the judges asked him for a demonstration he replied that he had already destroyed the magic branches. Grandfather and hired killer were both executed.

Human beings are also killed by real crocodiles, however. Two European children were bathing in one of the crystal-clear pools at Mzima Springs in Kenya's Tsavo Park. The girl saw a crocodile approaching under water and scrambled ashore, but the boy was seized. Having searched the entire area with the aid of volunteers, his parents found his comparatively unmarked body hidden deep in a clump of reeds. In 1917 the animal dealer Hermann Ruhe lost an extremely competent catcher named Kreth as a result of unusual behaviour on the part of a crocodile in Sumatra. Returning from a hunting-trip with some friends one evening, Kreth casually draped his bare leg over the side of the boat. A crocodile seized it and wrenched him into the water so quickly that nobody could get in a shot. His mangled body was found two days later, after the river had been searched by local troops.

Crocodiles in love

When mating, male crocodiles often emit a long, booming roar which sounds like the rolling of big drums. To do this, they raise their heads and open their mouths wide. Courtship display takes place in the water – at Lake Rudolf mainly between 9 and 11 a.m., after the first sun-bathe on shore. A male which meets a female while patrolling its particular stretch of shore arcs its tail out of the water so that the tip remains submerged. It also raises its head until the lower jaw lies on the surface, inflates its neck conspicuously, and sends bubbles rising from both cheeks. Having whipped up the water by clashing its jaws together and violently flailing its tail, it pursues the female, swimming alongside and eventually overhauling her until it forces her to swim in a circle. The male now remains silent, whereas the female raises her head from the water and utters a series of throaty cries. She either takes flight at this stage or permits her suitor to place one foot on her shoulder and mount her. The creatures' tails may become partly entwined and beat against one another. The male clings tightly to the female so that both crocodiles are able to swim along together. In deep water the female becomes submerged, in shallows the male lies on its side. Actual copulation may last from thirty seconds to almost two minutes. Mating behaviour ceases between 1 and 2 p.m.

When territory-owners become involved in fights with their neighbours or aggressive male outsiders, the two adversaries lie facing one another submerged with only the tops of their heads showing above the surface. Their nostrils send little fountains of water jetting into the air. Suddenly, one crocodile lunges at the other with jaws gaping and shoulders clear of the water, sometimes emitting a loud roar. The two rivals may exchange bites for up to three-quarters of an hour and occasionally lock jaws in the process. Other crocodiles are attracted to the scene.

Brooding-times are peculiar to individual areas and vary in different parts of Africa. On the Victoria Nile the crocodiles lay their eggs between December and January, or during the dry season when the water-level drops. On Central Island in Lake Rudolf, the eggs are laid at the end of December, after the first rains, and

Crocodile nests

the young hatch out after the second rains have begun in March.

Nesting-places are situated 5-10 yards from the water and about 6 feet above the surface in a well-shaded, flattish stretch of bank or shore where the sand is free from stones. The females dig with their feet, pushing the sand away with their hind legs and tearing up grass with their teeth and claws. The egg-chamber lies 7-20 inches deep in the damp sand, being shallower in well-shaded spots than in shadeless ones. It has a circular entrance in front and tapers towards the mouth of the big egg-chamber. The female lays at night and produces an average of 35 eggs in several batches. She then covers the nest with sand or a layer of grass which may be as much as 10 inches deep. This grass keeps the sand anything up to 10°C. cooler than the ground in the immediate vicinity. Modha has recorded temperatures of 30°C. (86F.) to 35°C. (95F.) in the egg-chamber itself. The average temperature inside 10 nests examined at Murchison Falls Park was 30°C. (86F.), and the variation over 24 hours never exceeded 3°C. Some nesting-places are as much as 30 yards from the water.

Incubation takes between 11 and 13 weeks, throughout which time the mother guards her clutch. In damp muddy ground eggs may be only 4 inches beneath the surface, in sand as much as 20 or 24 inches deep. The mother spends most of her time lying on them, only taking an occasional dip when the heat becomes intense. Her body moistens the ground above the eggs when she returns. In earlier times, crocodiles probably laid their eggs in breeding-colonies like many species of birds. Cott found 24 nests in an area measuring 670 square feet at the southern end of Lake Albert. The mother does not normally abandon her eggs for the water during the hottest hours, merely retires to the shade of a neighbouring tree and keeps an eye on them from there. Crocodiles become extremely sluggish while brooding, probably because their bodies are dehydrated by evaporation.

The Nile varan, a lizard measuring between 5 and 6 feet, is noted for its habit of digging up crocodiles' eggs. It strikes them against a stone or tree before devouring them.

Baby crocodiles chirp below ground

Young crocodiles can be heard chirping prior to hatching out, even through a 12-inch layer of sand and from a distance of 12 feet. They are particularly prone to 'grunt' in response to blows on the ground, passing footsteps or tape-recorded reproductions of their own cries. Voeltzkow noted in Madagascar that the young crocodiles he was keeping in a box 'quacked' inside the egg as soon as he stamped on the ground or knocked against their artificial nest. The little creatures probably respond to the footsteps of their mother, who has to scrape away the sand with her belly before they can reach the surface – something they could never do on their own.

Crocodiles are bred for re-stocking purposes at the Mkuzi Game Reserve in South Africa. The eggs collected there show remarkable powers of survival. Packed in straw, they have safely withstood trips of up to 150 miles by water or along poor roads. All that must be done is to mark the upper surface of each egg so that it can be correctly positioned for further incubation – not in sand but between layers of grass in baskets which are sprinkled with water and placed in the shade if necessary. More young are produced in this way than in the wild, where the mortality rate is very high. The correct temperature for incubation ranges between 27°C. (80.5F.) and 35°C. (95F.), though it can drop to 19°C. (66F.) or rise to 38°C. (100F.) for brief periods without fatal results.

Crocodiles hardly ever attack human beings on land, but mother-crocodiles will do so at hatching-time. Hans Besser was attacked on land by a crocodile in Tanganyika. He killed it and then examined the spot where it had previously been scraping away at the sand. Under the impression that he could hear sounds rising from below, he put his ear to the ground and distinctly heard a noise 'rather of the sort made by sucking puppies when they groan with pleasure as repletion sets in'. He scraped the sand away with his hand and soon brought to light several young crocodiles which immediately snapped at him. They were about 10 inches long, and some of them still had umbilical yolk-bubbles. They at once headed for the river, whose surface was invisible to them, and per-

Riding on mother's back

severed in their attempts to reach it even when Besser barred their path. He took 30 of them and bottled them up in a gourd, intending to bring them home alive. The rest he killed with the blade of an oar as they struggled to the surface. The egg-chamber contained a total of 205 young crocodiles. Some of the eggs were still unbroken, but the lime-toughened skins were easily ruptured. The young crocodiles then burst the sac themselves and made for the water with yolk trailing behind them. All the eggs were good, but Besser could not determine whether they had been laid by one female or several. He could discern absolutely no signs of growth in the young crocodiles he took with him despite three years' generous feeding on minced meat and small fry. The little community always fell on its food and ate under water, never on land.

Young crocodiles are at the mercy of hawks, marabous and vultures during their first few hours of life. The Nile varan pursues them into the water, where they are additionally preyed upon by turtles and adult crocodiles. The mother-crocodile escorts her young like a duck, even allowing them to climb on her back. Baby crocodiles are 11-13 inches long at birth and have a yolk-sac the size of a hen's egg which contains reserves of nourishment sufficient for several months. They stick closely together at first, indefatigably squeaking, grunting and snapping at each other, crawling into nooks and crannies, greeting their mother with a chorus of grunts, clambering over her head and snout, and promptly swimming after her when she dives. Mothers drive varans, other crocodiles and herons away from their nurseries, but losses are very great. At Lake Rudolf, large numbers of young crocodiles are taken by catfish as well.

In January 1968, while visiting the crocodile's southernmost African breeding-ground beside Lake St Lucia in Zululand, South-East Africa, A. Pooley found that 22 nests out of 65 had been completely ravaged and destroyed by Nile varans – and this at the very start of the brooding-season. Varans are bold and determined thieves who wait patiently for the right moment to plunder a nest. One was observed removing an egg from a nest no more than 20 feet from the spot where a large crocodile was guarding a nest of its own. Another big varan cautiously crept between two crocodiles sunning themselves only 20 feet apart, discovered an unguarded nest, and proceeded to dig it up. Yet another varan took an egg from a nest beside the Pongola river while two crocodiles were guarding their nests no more than 10 yards away and the unwitting mother lay in the river only 5 yards away. Varans always adopt the same procedure. Having found an unguarded nest they dig until they uncover the topmost layer of eggs, take one egg only and carry it in their jaws to the nearest cover, which may be 10 yards away. Once the egg has been devoured a second is purloined and borne off, though not necessarily to the same hiding-place. This goes on until the thief has had enough or is disturbed.

Young crocodiles' prospects of survival are greatest when all the females begin to lay their eggs at the same time. If they start to lay at intervals of several days, there is a strong risk that each nest will be robbed in turn.

Another creature which preys on crocodile eggs at Lake St Lucia is the marsh mongoose (Atilax paulinosus). This opens the nest, removes an egg, and bites a neat hole in it. The liquid content of the egg runs off, leaving the embryo behind. Marsh mongooses operate mainly at night. The turtle Trionyx triunguis takes a heavy toll of any young crocodiles that do manage to reach the water, and in Zululand the rusty-spotted genet (Genetta rubiginosa) is a notorious nocturnal predator of baby crocodiles. The African eagle-owl preys regularly upon them too. Of 838 crocodiles collected by Pooley from 25 nests, 12 died of umbilical infection in the first month after hatching. Another 21 were weaklings, and far below average size. The incubation period in Zululand varies between 12 and 14 weeks. Modha ascertained by marking them that members of the same brood remain in the same spot for two or three weeks before joining forces with young from other nests. Females guard other females' young as well as their own.

Young crocodiles live concealed behind undergrowth in weed-choked inlets, in other words, precisely where their elders do not

Undisturbed by the camera

live. They begin by preying upon snails, dragon-fly larvae, crickets and other insects, then crabs, toads, frogs, small birds, and rodents. Half-grown crocodiles live on fish and snails, but as time goes by their diet comes to include more and more mammals and other reptiles.

Joy Adamson, celebrated foster-mother of Elsa the lioness, found seven baby crocodiles in a small pool not far from her tent. Thanks to their protective coloration and large black patches, they were scarcely distinguishable from their surroundings.

Their heads seemed unusually large in relation to their bodies; their eyes, which corresponded in size to large peas, were a pale ochre and protected by heavy brow-protuberances. They often clung to floating reed-stems and swam by vigorously kicking and treading water. They dived immediately at the slightest movement, even when the source was 6 or 7 yards away. On the other hand, the sound of human voices or a camera clicking did not worry them in the least. They took as little notice of morsels of meat on the end of a stick as of worms, dragon-flies and flies dropped in their midst or circulating in their immediate field of vision, but when Joy Adamson's husband George imitated the 'imn, imn' of a crocodile they promptly closed ranks and turned their heads in the direction of the sound. They did not, however, come any closer and continued to hug the reeds. This at least proved that they were not deaf. Voices and camera-clicks must therefore have been heard by them but dismissed as meaningless.

Warmth and shade are considered particularly important by those who rear crocodiles artificially at the Mkuzi Reserve in Zululand. Young crocodiles must be enclosed by $\frac{1}{2}$-inch mesh or they will climb through. Fencing must also be dug to a depth of 18 inches and the interior lined with slate to prevent them from burrowing beneath it. A height of 3 feet above ground is sufficient to contain them provided the wire mesh curves inwards. The inner fence at Mkuzi is surrounded by a dense hedge composed of wooden posts and reeds designed to give protection from wind and predators. The man-made pond at their disposal is divided in half by a small causeway. One half contains cold water and the other is used for the introduction of heated water during the cold season so that the creatures will continue to feed. An electric light suspended above the surface acts as an insect-trap. If young crocodiles are fed on meat they have to be given additional calcium. At the end of their second year they are released into rivers. The young crocodiles double in size during their first year. By the end of their first year they are about 30 inches long. The crocodile's average annual rate of growth in the first 7 years is $10\frac{1}{2}$ inches, but this drops to only 1.4 inches by the time it has attained 22 years.

Even when at liberty, half-grown crocodiles shun the world. As Pitman wonderingly remarks, crocodiles between 2 and 5 feet long do not seem to exist. In fact, specimens of this size are sometimes encountered far from the nearest water. It is evident that they avoid the company of larger crocodiles for fear of cannibalism. A. C. Pooley while in Zululand during March 1968 seems to have hit upon the real reason why one hardly ever sees young crocodiles of a certain length. The Natal Park authorities collect crocodile eggs there, incubate them artificially to avoid the heavy losses normally inflicted by predators and rear the young crocodiles in man-made basins. These concrete basins were supplemented by the temporary expedient of sinking unlined pools into the ground. After a week of cold weather it became clear that ten crocodiles 15 to 20 inches in length had dug a tunnel into the side of their pool just above water-level. The tunnel, which was two feet long, led into a larger cavity in which the crocodiles lay. 60 yearlings in another earth-lined pool did precisely the same thing. Although they had been there for more than three months they did not begin to dig until a period of cold weather set in. These two- to three-footers dug twelve burrows in all and abstained from food entirely once they had started operations. They all remained in their burrows unless it was warm and sunny, but only the biggest of them ventured to bask outside on sunny days.

The largest of the tunnels was 13 feet long, whereas the smallest varied between 4 and 6 feet in length. The passages ran straight and

A 19-foot specimen

horizontally. If the water in the pool had risen by only 4 inches, the burrows would have been completely flooded. When dug up, all crocodiles were found with their heads facing the entrance. They made no attempt to bite or run away, merely dug their claws passively into the ground and braced their bodies against the earth walls. The chamber at the end of each tunnel contained five or six crocodiles packed tightly together two or three deep.

It was at first assumed that the crocodiles started work on their burrows with their claws, but it eventually transpired that they bit into the bank just above the surface where the soil was soft. Then, with earth trapped between their jaws, they turned, dived, and shook their heads to and fro, presumably with their mouths open. Two or three crocodiles could be seen working on the same hole.

June and July in the southern hemisphere brought Zululand its first frost for five years. Many of the young crocodiles kept in concrete basins succumbed to the cold or contracted pulmonary diseases. There was no death or sickness among the young crocodiles in earth-lined pools. They had all dug themselves into the bank within a week, five to a burrow on an average. Each tunnel was installed in the south or south-east side of the pool. This shielded its occupants from the prevailing cold wind and rain, which come from this direction, whereas the north winds, which are warm, blew straight into the mouth of each burrow. All the young burrowers resumed feeding in September, when the weather turned warmer. The little crocodiles vanished into their holes immediately at the first sign of danger. Even during the hot season they preferred to remain in their shady burrow rather than in open water where they were at the mercy of predators, especially from the air.

Male crocodiles do not become sexually mature until they attain a length of $9\frac{1}{2}$-11 feet, females at 8-9 feet, so they must be at least 18 years old. It takes a very long time to re-stock an area with crocodiles even when the strictest protection is enforced.

The Uganda game authorities reported the killing of a crocodile on the Semliki Nile measuring $19\frac{1}{2}$ feet in length and 5 feet 8 inches round the belly. In March 1903 Hans Besser claimed to have killed a 25-foot specimen which had lost one-quarter of its tail. It was 14 feet round the middle and its skull was 4 feet 7 inches long and 37 inches wide.

Monsters such as these have always whetted the curiosity of the European. M. Aemilius Scaurus was the first to exhibit crocodiles at Rome in 58 B.C. – five of them in a specially dug pool. When dedicating the temple of Mars Ultor in 2 B.C., Emperor Augustus had water channelled into the Flaminian Circus and decreed a public entertainment at which 36 crocodiles were slaughtered. Several more such spectacles were staged later. Emperor Elagabalus kept a pet crocodile in his palace, and crocodiles have been favourite circus exhibits in recent decades. Their mouths are sometimes tied with thin, almost invisible cords to prevent them from opening their jaws and biting. The same method is employed in zoos when crocodiles are to be caught and transported or when large specimens must be induced to share the same quarters without immediately biting each other. (They can, after all, go without food for a considerable time!) Crocodiles can also become quite tame – so tame that they let themselves be carried around during circus acts with their toothy jaws wide open. These, however, are usually American alligators which never attack human beings in the wild.

11. Animals and fire

Anyone who fishes a fly out of his food is not hungry.

AFRICAN PROVERB

At 6 a.m. on April 8th 1896 a man was found lying in the bear-pit at Berne Zoo, his body stripped of everything save shoes and socks. Mani, a sixteen-year-old brown bear had spread its paws over the corpse and would not let its mate come near. The victim, a young man who smelled strongly of drink, had evidently tried to bait the animals during the small hours. He must have leaned over too far in his befuddled state, lost his balance, and fallen. Bundles of blazing straw were thrown into the pit in an attempt to drive the bear away, but it beat out the flames three times in succession and dipped its paws in water afterwards to cool them.

It would certainly be incorrect to say that wild animals always panic at the sight of fire. When a lamp exploded one evening in Joy Adamson's camp, the cubs belonging to her celebrated tame lioness Elsa continued to lie there calmly, eyeing the pillar of flame with mild surprise. Their mother went so close that she had to be warned not to singe her whiskers. When bush-fires sweep the Serengeti Plain, the lions sometimes bask in their heat or roll in the fresh ash while it is still warm. Zebras, antelopes, elephants and Cape buffalo not only refrain from racing off in panic, contrary to what one reads in so many books about Africa, but often remain in close proximity to the flames. What is more, fire holds a powerful attraction for all birds which live on insects, lizards, frogs, snakes and other small creatures. Storks, marabous, bustards, eagles and buzzards all flock to the scene from far and wide to hunt the lesser game that have to flee from the flames.

The inquisitive elk

Fires started by lightning are extremely rare in Africa because thunderstorms are usually accompanied by falls of rain which quickly extinguish the flames. Wild animals do not run from thunderstorms, however noisy, even though they are as vulnerable to lightning as we are. On March 22nd 1964 seven deer were simultaneously struck by lightning and killed at Ramerberg am Inn. The act of fleeing in blind panic only diminishes a creature's prospects of survival. Not only is it impossible to escape from lightning, but panic-stricken fugitives are only too liable to be seized by predators. Shortly after the war, when the zoological gardens at Frankfurt were completely devastated and almost stripped of animals, we raised money for their reconstruction by organizing a variety of public entertainments. We even put on firework displays, having carefully assured ourselves in advance that our charges would be quite untroubled by the attendant flashes and bangs.

Curiosity is not peculiar to man. In Europe, as opposed to Africa, fires are comparatively rare in the wild. Free-ranging animals may be startled and put to flight by the unwonted spectacle of yellow flames. On the other hand, they may be irresistibly attracted. Early one morning, when Baron von Tiesenhausen was stalking grouse on his estate, an elk inquisitively approached his recumbent form, snorted at the unaccustomed sight, and stamped its hoofs. The Baron lit a match in the hope of frightening it away without undue noise. The elk galloped off but was soon overcome by curiosity and returned. It came right up to the Baron and bellowed in his face – a sound calculated to strike terror into the stoutest heart. Tiesenhausen now lit almost all his remaining matches simultaneously and threw them into the elk's face. Only then did he succeed in putting the big animal to flight. Similarly, fishermen and expedition porters have more than once been attacked by hippos on islands in the Lulua river in the Congo while sitting or sleeping round camp fires.

Elisabeth Warmbrunn of Hamburg describes her dog's first experience of Christmas as follows: 'Then Nero came in, caught sight

The dog that blew out candles

of the candles on the Christmas-tree, which was standing on the ground, and sniffed one of them. The thing "bit", so had to be bitten in return. Another pain was the signal for him to bark for the first time. "Woof!" – and out went the light. Quickly on to the next, and another "woof!" It didn't work with the third candle, which went on burning. Very gingerly, a paw was enlisted to snuff out the nasty "animal", and so on until every candle within reach had been barked or snuffed out. From now on the dog used the same technique to extinguish lighted matches whenever they were held out, a trick which delighted customers in the bar. One of them threw the dog a smouldering cigar-butt to see what would happen. Nero snapped at the butt, burned himself, and dropped it. Having thrown himself on the ground and rolled around on the unfamiliar "animal" until it was dead, he promptly swallowed it. Hard to tell how many cigar-butts and cigarette-ends Nero "smoked" in this way, but he became sick – so sick that we were careful never to take him into the public bar again.'

Redbreasts and starlings often set fire to their nests or nesting-boxes with burning cigarette-ends or cigars which they have introduced into them as building-material. A tame redbreast which had been startled by a bird of prey swooping past the window shot through the open door of a kitchen range but promptly flew out again unscathed. Why do so many crows and jackdaws display an almost morbid-seeming passion for flames and smoke? Tame birds have learnt how to strike matches without being taught. They hold them beneath their wings and positively bask in the smoke and flames, hardly singeing or injuring their plumage at all in the process. Nobody has yet discovered the point of this behaviour. Similarly, many birds will pick burning wisps of straw or twigs out of a fire and fly on with them for some distance. It was repeatedly alleged in earlier times that thatched roofs had been set on fire by birds. In the year 1201, for example, birds were said to have started a big fire in London by this means. Recent observations suggest that such reports should not be dismissed too lightly.

Needless to say, very convincing films can be made of huge

The day the sun changed colour

herds of animals stampeding in the path of bush-fires. All that is required is to chase zebras, gnus and antelopes across the plain by car, film them at full gallop, and intersperse these sequences with others showing expanses of grass-land going up in flames. Anyone looking at the finished product will assume that the animals are fleeing in terror from the fire. The terrible forest fires which really do kill large animals and human beings occur in Europe, Asia and North America – in other words, in more temperate regions. The jungles of Africa and the tropics are too damp to burn readily. In 1955, California suffered no fewer than 436 forest fires during an eighteen-day period spanning late August and September, and there were days when as many as 14,000 people were engaged in trying to control them. In the same year, 154,160 fires were reported in the United States as a whole, and that was the lowest figure for a decade. In Canada, thirteen forestry officials were burned to death in a single year after being dropped by parachute and supplied by air in a bid to nip fresh outbreaks in the bud.

Between September 26th and 28th 1950, the sun and moon looked quite blue to observers at widely scattered points in Germany, England, France, Switzerland and Norway. The sun was pure white when it set, not red. In the United States it was so overcast by day that people could stare into the eye of the sun without being dazzled. What had caused this discoloration, in Europe as in America, was that more than a hundred forest-fires were raging simultaneously in Western Canada. Their smoke had travelled half-way round the globe. Finally, on September 27th, they were extinguished by rain and snow.

In Africa, fire has clearly been instrumental in the spread of sleeping-sickness. Like the related cattle disease known as nagana, this is transmitted by the tsetse fly. Recent research has shown that the tsetse fly lives mainly on the blood of the wart-hog and derives far less of its nourishment from other forms of game. A widely used method of combating nagana among cattle in the old days was to make life impossible for the tsetse fly by systematically shooting all wild animals within a wide radius. Unlike zebra, buf-

Fire and tsetse flies

falo, giraffe, elephant and many species of antelope, wart-hogs vanish into their labyrinthine burrows at the least sign of danger and are virtually immune to mass-extermination by such means. Working in Zambia, formerly Northern Rhodesia, the biologist B. L. Mitchell points out that wart-hogs feed almost exclusively on grass, grass-roots and one or two varieties of sedge and reeds. In the highlands they live almost entirely on the grass known as Loudetia superba and the roots of various Hypanhenia species. These grasses predominate on the high plateaux.

Forestry officials in Northern Rhodesia carried out experiments with fire over a period of 25 years. It emerged that severe annual burning causes woody plants to disappear and encourages the growth of tough, fire-resistant grasses such as Loudetia superba and Hypanhenia. The increased incidence of fire in the present century may thus be presumed to have created conditions favourable to an increase in the numbers of wart-hogs and, consequently, of tsetse flies. This may well have contributed to the spread of the tsetse in the last sixty years.

Grass is burnt off nearly everywhere in Africa to facilitate hunting. Fires are started every year in places where wild animals are shot in an attempt to control the tsetse fly. As far as I can see, this form of anti-tsetse drive only produces more wart-hogs and more tsetse flies. It is also to be hoped that the authorities responsible for these mass-shootings will finally stop exterminating the Bubal Hartebeest, which never seems to play host to the tsetse.

I do not mean to say that fire never kills large animals in Africa. In many years the rainy season brings unusually heavy precipitations. In 1962-3, for instance, East Africa had more rain than in any of the preceding forty years. Bridges were washed away and roads destroyed. Thanks to the intense humidity, the grass in many areas grew taller than a man and extremely dense. It was easy to picture what would happen when it turned parched and yellow in the ensuing dry season and twenty-foot walls of flame swept through it.

Fires are not only unavoidable in Africa – man has deliberately

Fire-lanes and their benefits

kindled them year after year for the past millenium or more. The fact that the Sahara creeps southwards and the Kalahari northwards every year, in other words, that the continent is steadily turning into desert, may well be due to the persistence with which man lights these grass-fires.

Using large earth-moving machines in 1963, we quickly protected the most vulnerable areas of Tsavo National Park in Kenya with 110 miles of fire-lanes. The grass between two parallel cuttings was burnt while still half-green, creating a bare strip which effectively halted the big fires that broke out at the height of the dry season and prevented them from entering the game reserve. No animals died an agonizing death there thanks to these prompt measures, which were, incidentally, financed by contributions from my television audiences in Germany.

Unfortunately, men who are out for personal gain have never had much consideration for nature or the future of their own children. All the Mediterranean countries which used to be pleasantly wooded before the golden age of antiquity – from Palestine, Turkey and Greece to Italy, Spain and North Africa – were stripped of their forests to build the wooden fleets of the Greeks, Romans, Turks and Spaniards and fuel the many bathhouses of the ancient world. Today, large tracts of Spain, Yugoslavia and Greece are barren and cheerless. The fertile soil that clothed their mountains has been washed away, and reports of terrible floods are commonplace. Of all the inhabitants of the Mediterranean area, only the Israelis have, by dint of great effort, artificially re-afforested their mountains on a large scale. Even in Australia, which is certainly an arid continent and becoming progressively more so, the few areas of forest have been destroyed because wool fetches such high prices on the international market. Sheep farmers ring the bark of large trees until they wither and die, then fire them so as to be able to graze huge herds of sheep on the land in due course. Dr H. O. Wagner, a good friend of mine who drove by car from Brisbane to Sydney in January 1937, did not come across a single patch of forest or bush which was not burning, and saw twice as many charred trees as living ones. On a flight from Melbourne to Sydney in the following year he counted seventeen forest-fires from the air. So many millions of attractive little koala bears were annihilated in this way that the few survivors are now carefully preserved in special sanctuaries. The position had not improved when I myself toured Australia by car thirty years later.

In 1965 a fierce bush-fire broke out in the Longwoo area, about 80 miles north of Melbourne. Thousands of animals perished in a sea of flames which also claimed eight human lives and devastated 100 square miles of hinterland. The fire broke out at a time when Victoria was experiencing its hottest weather for many years. The air temperature reached 43.3°C. (109.8F.) Policemen and soldiers armed with automatic weapons mowed down at least 6,000 sheep and cattle which had been marooned by a raging inferno so as to spare them an agonizing death by fire.

In Africa, even elephants occasionally perish as a result of fires. In 1956 natives in Uganda killed 64 elephants out of a single herd by deliberately encircling them with flame. Might not such incidents account for the ineradicable myth of the elephants' graveyard? The discovery of such large concentrations of elephants' bones in the bush could well have prompted fanciful explanations of this kind. In 1958 a bush-fire burnt to death 15 young elephants, some of them only two years old, in Tsavo National Park. The carcases were carefully examined to determine whether they had died previously of some disease or been weakened by it. Five more elephants were found to have been killed by fire at another spot, and the areas involved are so vast that there may well have been numerous other cases.

The noted South African snake expert F. W. Fitzsimmons relates how an entire village was threatened by a blaze which the inhabitants had themselves started. They fled to a bare rocky hillock but were more than a little alarmed when large numbers of animals joined them there, together with pythons and many poisonous snakes such as mambas, puff-adders and cobras. The creatures were so bemused and disarmed by their unusual predicament that no one came to any harm. A similar situation arose in England in 1958 when Manchester Zoo's aviary went up in a sea of flames 50 or 60 feet high. As soon as the fire broke out the director of the zoo and his keepers rushed into the building to save the birds. They opened the cages and grabbed any birds they could, then threw them indiscriminately into the handful of spare cages that were housed in other buildings. Laughing geese were so overcome with agitation that they emitted their weird cries again and again, parrots screamed and repeated the words they had been taught. Species of all kinds, including birds of prey, were cooped up together, but there was no mutual hostility.

The effects of a circus fire tend to be worse because large circus animals are generally chained or tethered. Apart from that, tents blaze up very quickly and collapse on the immobile animals beneath. In 1931 the Sarrasani Circus from Germany paid its first postwar visit to the Belgian city of Liège. In the afternoon a parade marched through the streets, and that night the whole circus suddenly went up in flames. Damage was estimated at over a million marks, and 8 out of 22 elephants burnt to death. One of them plunged into a moat in its agony, broke the ice and injured its spine so badly that it could not be extricated.

At Cleveland, Ohio, on August 4th 1942, the menagerie tent of Barnum and Bailey's Circus caught fire and burnt down in three minutes. 65 animals died in the blaze or sustained such terrible burns that they had to be shot, including 4 elephants, 13 camels, 12 zebras, 2 giraffes, 4 lions, 3 tigers, 16 monkeys, and a zebu. The elephants were roasted alive, the flesh in some cases peeling off their bodies in broad strips a yard long, and several of them had their ears almost completely burnt off by the flames. The zebras behaved so uncontrollably that there was nothing to be done with them, but the elephants did not move until their trainer appeared. At a word from him, each elephant seized its stake with its trunk and wrenched it out of the ground, grasped the animal in front by the tail, and marched out in good order. The circus veterinary surgeon, J. Henderson, wrote afterwards: 'I realized then that there must be a characteristic in animals comparable with the inner greatness of the human being. It is not their physical size alone that counts, nor their agility, wildness or strength – no, they possess a kind of spiritual nobility, an inner relationship to that which is enduring in Nature. At that moment I acquired a respect for them which I had never known before.'

When a truck laden with explosives caught fire and blew up at a snake-farm in Marshalls Creek, Pennsylvania, six people were killed and ten wounded. The explosion tore a hole in the ground 4 feet deep and 30 feet across. Some time afterwards it was discovered that hundreds of poisonous snakes had escaped. A large team of policemen and firemen went to work on the dangerous fugitives with sticks and fire-arms. At least forty were destroyed, including a number of cobras.

The naturalist William Beebe once watched molten lava flowing into the sea during a volcanic eruption in the Galapagos Islands. Sea-birds, mainly petrels and divers, were attracted by the dead fish, but some of them paid dearly for their greed. Saddest of all, Beebe wrote, was the fate of a full-grown sea-lion which suddenly sprang into the air in the immediate vicinity of the shore. Five times it soared above the seething water in an arc 8 or 10 feet high, and then, blind with pain, headed straight for the red lava delta. No death-struggle was observed. With a final bound, it leapt into the jaws of death.

Many cattle breeders used to believe that domestic animals could be cleansed from or immunized against disease by fire. This fire had to be specially kindled by rubbing one piece of wood against another. There is evidence of this practice in Central Europe as far back as the 8th century, but countryfolk in England and Scotland were still trying to cure cattle by applying 'virgin' fire in the 1840s. In spring 1855, villagers in Brunswick drove diseased pigs through a fire which the local mayor had kindled in accordance with ancient tradition. Pigs were also driven three times through a large bonfire at Gandersheim earlier in the 19th century. Fires were extinguished in every house, later to be rekindled with burning logs from the healing blaze. Peasants also believed that fire could protect their livestock against witchcraft.

Human beings living in the wild can barely survive without fire. In 1906 ten shipwrecked sailors took refuge on Indefatigable Island, one of the Galapagos group in the Pacific. Having lived for two-and-a-half months on the blood and raw flesh of giant tortoises, they discovered to their annoyance, shortly before being rescued, that one of them had been carrying a box of matches in his bag the whole time without knowing it.

We usually underestimate what human beings can endure in the way of heat. In the last century, demonstrations were staged in several European countries by a performer named Ivan Ivanitz Chabert, who claimed to be 'an incombustible phenomenon'. He had a huge metal oven heated on stage, then went inside and remained there until a leg of mutton suspended from the roof was thoroughly roasted, also some pieces of steak on a plate which he was holding. According to contemporary accounts dating from 1867, he withstood a temperature of 195°C. (383F.) and even sang inside the oven.

Although variety turns of this type tend to be regarded as superhuman feats or cunning pieces of deception, they all depend on entirely natural processes which anyone can test for himself. Scientists belonging to the Royal Society had already made this discovery in London in February 1774, though without causing any great public sensation. Dr Charles Blagden, a respected scientist who was later knighted for his contributions to medicine and oceanography, demonstrated similar experiments to members of the Society. In company with three colleagues, he spent some time in a chamber which had been heated to the bread-oven temperature of 120°C. (248F.) by means of a centrally located iron furnace. They at first entered together but discovered that their own bodies caused a rapid drop in temperature. From then on, only one person entered the chamber at a time. When Dr Daniel Charles Solander, a friend and pupil of the great botanist Linnaeus and participant in expeditions to the Antarctic and Iceland, went in, the temperature fell in three minutes from 100°C. (212F.) to 92°C. (197.5F.) Dr Blagden described how the ability of the human body to maintain its natural temperature gave rise to some surprising phenomena. Whenever he and his colleagues breathed on a thermometer the mercury dropped several degrees. Each exhalation produced an agreeably cool sensation in the nostrils, which had been scorched by the hot air inhaled a moment before. In the same way, their now cold breath cooled their fingers pleasantly. When Dr Blagden touched his side it felt as cold as that of a corpse. His actual body-temperature, measured under the tongue, stood at 37°C. (98.5F.), or about one-tenth of a degree above normal. Men who were experiencing no distress in air heated to 100°C. (212F.) (boiling-point) found it unbearable to touch mercury standing at 51°C. (123.5F) and could only just endure contact with spirits of wine at 55°C. (131F.) All metal objects, even their watch-chains, were so hot that they could hardly bear to touch them even for an instant, whereas the air from which the metal derived all its heat was merely unpleasant . . .

These bake-oven temperatures evidently did the scientists no harm, because Sir Joseph Banks was President of the Royal Society from 1778 to 1820 and died at the age of 77. Dr Blagden conceded that he and his companions did experience some discomfort at times. Their hands trembled a good deal and they suffered to a considerable extent from exhaustion and debility. Blagden himself experienced dizziness and a roaring in the ears. Their clothing, he said, protected them as much from the heat as it would have done from cold. Beneath it they were surrounded by air which had been cooled to 37°C. (98.5F.) by contact with their bodies and was only gradually heated from the outside because wool was such a poor conductor. The scientists came straight out into the open air without taking any precautionary measures, but suffered no ill effects. Exhaustion and the trembling in their hands soon disappeared.

During a later experiment the scientists exposed themselves to a temperature of 130°C. (266F.) Their pulse-rate almost doubled

The lightning speed of the lion

to 144 per minute. Blagden also entered the chamber stripped to the waist. The initial impact of the hot air on his naked body was, he said, far more unpleasant than anything he had experienced through clothing, but within five or six minutes he broke out in a sweat which brought him immediate relief. Only the ability to sweat keeps a living body cool. The bare-chested scientists managed to survive the inferno whereas some eggs and pieces of meat exposed to the same air temperature were cooked within 13-47 minutes. The air in sauna baths is heated to only 50-56°C. (122-132.7F.) Human beings have a special ability to endure heat because they probably possess more sweat-glands than any other mammal and are therefore supremely well equipped to disperse body-warmth.

A humble cigarette-end provided me with a striking demonstration of the lightning speed with which predators can switch from repose to action – probably their most effective predatorial 'trick'. It happened at Frankfurt Zoo, less than a year after the war ended. A young American soldier crossed the safety barrier in front of the lions, which were still kept in cages as opposed to the open-air paddocks of today. One lion was lying so close to the front of the cage that its tail hung down through the bars. The soldier stubbed out his cigarette on the base of the animal's tail. Almost simultaneously the lion swung round and ripped off the man's scalp, pulling it down over his face. Scalps heal easily, so the soldier was hurried to hospital and sewn up again. As far as I know, he recovered without suffering any serious after-effects.

12. *Escapade in the Sudan*

The man who tries to storm the world at a zebra's gallop will end at the pace of a chameleon.

AFRICAN PROVERB

The whole affair began quite innocently.

After testing my Amphicar in Uganda, I decided to make the most of my last few days in Africa and visit Lake Rudolf. Situated in Northern Kenya, this was one of the last big African lakes to be explored. It was discovered on March 6th 1888 by the wealthy Hungarian nobleman Count Samuel Teleki von Szek and his companion Lieutenant von Höhnel after a fourteen-months' march and a host of adventures and skirmishes.

Seventy-seven years later, I made things rather easier for myself by hiring a light aircraft in Kampala, the capital of Uganda, which lies at the northern end of Lake Victoria. These air-taxis – mostly single-engined four-seaters – not only get you to your destination far more quickly in East Africa; they are far cheaper as well. Making a bee-line across land and water is far more economical of mileage than using the scanty roads. I was accompanied into the light and rather fragile-looking plane by my friend Alan Root. The distance between Entebbe, Kampala's airport, and the western shores of Lake Rudolf is roughly 300 miles. The last hour-and-a-half took us across semi-desert and mountains – broad uninhabited tracts of gleaming yellow-brown broken by an occasional patch of green. This was the land of the Turkana, a tribe of nomadic cattle- and sheep-breeders whose menfolk still go around naked today.

Twice, three times, our little machine circled the lake-shore until the British pilot decided to regard a sprinkling of sticks and

Toast to a Crown Prince

white stones on an expanse of sand as a marked landing strip. We stirred up a huge plume of dust as we rolled to a stop. Around us were a few palms, an emaciated dromedary wandering alone, and sand, sand, sand.

Bare-bodied young men picked up our baggage and carried it across to a large straw hut beside the water. This was the home of a lone European, a Briton whose job it was to teach the nomadic Turkana how to fish in the lake. A lonely life, to say the least. That evening we went bathing at a spot where there were supposed to be no crocodiles. The sun slid over the lip of the cloudless blue sky, took a last look at its reflection in the equally blue and boundless waters of the lake, and disappeared behind the barren, greyish-red mountains. The water was warm and brackish.

When Count Teleki named the huge lake after the heir to the Austrian throne his supplies of wine and brandy had already run out in the course of his long march. He duly mixed brackish water with tartaric acid, carbonic acid and honey to make a ceremonial beverage which the explorers drained after raising three cheers to the crown prince whose name the lake now bore.

Returning barefoot across the yellow sand to our host's hut, I felt a persistent pricking sensation in my feet. It came from the bones of countless fish whose huge desiccated heads lay strewn in the grass. The tufts of grass were just as prickly. It was difficult to conceive how cattle and goats could crop and chew the hard spiny stuff, let alone digest it. They certainly looked thin enough.

Starvation is the main worry here. The Turkana, who have never been numerous, used to roam the huge area with their herds. A few years back there was a drought of the kind which afflicts East Africa once every thirty or forty years. Many people used to die in the old days, especially in these regions of semi-desert, but under colonial rule attempts were made to help. This time, funds had been sent by U.N.O. and supplies of food distributed in a camp at the south-west end of the lake. Food was still being doled out now, three years later. Little by little, people had drifted to the area and stayed there to raise families, living at subsistence level but never

Life on the dole

moving on. They are becoming increasingly estranged from their fellow-tribesmen, the nomads of the interior. Some 4,000 of them are still being fed there, and one shudders to think what will happen if the UN subsidies cease.

In the days that followed, days of perpetual sun and wind, we toured the lake in our fishing instructor's motor-boat. Anyone who does this will quickly develop a nostalgia for Lake Rudolf. The vast sheet of water is remote and set in an arid landscape, but the air has a pellucid clarity. 160 miles long, Lake Rudolf is fifteen times as big as Lake Constance and would stretch from London to Manchester. It has risen in the last few years and become much larger, like most of the East African lakes. We glided past skeletal palm-trees with cormorants and pelicans perched on them. These trees used once upon a time to be on dry land, and our fisherman friend's original house stood among them.

We flew along the lake-shore, half over land, half over water. Dozens of crocodiles deserted their sand-spits and slid hurriedly into the water. There were a few hippos, too, as well as hosts of

Page 193:
Male members of the Turkana tribe, which inhabits the remote region bordering Lake Rudolf, often go naked, especially when fishing. The same applies in the Southern Sudan, where I had to make a forced landing. Men in the villages there often came out to greet me wearing no clothes whatsoever. I was at first locked up by the Sudanese government and threatened with criminal proceedings.

Nile crocodiles exceeding 18 feet in length must be at least a century old. The creatures do not become sexually mature until 18 years of age. They always avoid human beings and large animals on land except when lying on a nestful of buried eggs, in which case they defend them. Little crocodiles can be heard 'chirping' while still in their eggs below ground.

Page 194:
In order to drink, giraffes have to splay their legs and sometimes bend them as well. Repeated scientific studies have been made in recent years of the way in which they cope with the resultant fluctuations in cerebral blood-pressure.

Page 195:
The suricates of South Africa, also known as grey meerkats, live in groups, the babies of the genus being cared for by their fathers, uncles and aunts as well as their mothers. The little creatures have extremely good eyesight and colour-recognition, and are for ever rising inquisitively on their hind legs. Being universally liked, they sometimes live on farms and even get on with dogs.

oryx antelopes, herds of dromedaries apparently unescorted by human beings, zebras, and Cape eland. Totting up the score in Count Teleki's diaries, one finds that he shot 81 rhinos, 2 lions, 75 Cape buffalo, 31 elephants, 162 antelopes, zebras, etcetera, not to mention birds, monkeys and numerous creatures which he only wounded. It should be remembered, of course, that he had to feed his porters. There are nothing like as many animals in the area today.

The northern end of Lake Rudolf extends into Ethiopia, and the uninhabited 300 miles of territory between its north-west corner and the borders of the Somali Republic are virtually regarded as an unpacified no-man's-land. Permission to go there is hard to obtain unless one travels in a well organized and well armed convoy of field-cars. It is only too easy to be ambushed and killed by roaming bands of Somalis.

On a small green patch of shore at the foot of a bare reddish-grey mountain we found a friend of Alan's, a young biologist, camping quite alone except for two Africans. The landscape is so uncluttered that signs of human activity are easy to spot from the air. The young scientist was not camping there voluntarily, and he found it unnerving that lions should venture so close to his little camp at night. He had actually been touring the lake in a metal-

Page 196:
Early European castaways described the Zulus as an extremely proud, clean tribe which prepared its food with great care and was very jealous of its womenfolk. It was not, however, until 1820 that Shaka, the Napoleon or Attila of South Africa, carved out his huge empire. He formed regiments of warriors who were not permitted to marry, made them go barefoot and abandon their spears, and trained them in hand-to-hand combat with short sword and shield. His empire took only a few years to build. One morning his palace was found spattered with blood. The witch-doctors who had established a reign of terror over his people were instructed, in accordance with ancient tradition, to summon a mass assembly and 'sniff out' the hundreds of guilty, i.e. bewitched, men who were supposed to have committed the crime. Ancient custom prescribed an agonizing death for these bewitched and trembling warriors, together with their entire families. Then, in booming tones, King Shaka proclaimed that he himself had sprinkled his house with blood. He at once decreed the execution of all the sorcerers who had marked out innocent men for punishment. Shaka was murdered by his half-brothers in 1828. His empire continued to flourish until 1879, when the British bloodily overthrew it after suffering heavy casualties themselves.

An air-drop of paperbacks

hulled outboard-powered boat to count and observe crocodiles. The Game Department of Kenya was anxious to know if crocodiles could be shot with a good conscience or if most of them had been exterminated, here as elsewhere, by poachers.

To cut a long story short, the crocodile-observer was driven on to a reef by high winds and heavy seas. His boat sank not far from shore at a spot where the water happened to be shallow. Although he and his companions managed to drag it ashore, the engine would not re-start after its soaking.

We duly flew back to the fishing instructor. It would be five or six days before he could cross the lake in a large motor-boat and rescue the stranded party, so we flew on to the southern end of the lake. Here stood a tourist hotel with pleasant chalets and a blue swimming-pool constantly replenished by a hot volcanic spring. (The proprietors, a plump Italian couple, together with their staff, were murdered two weeks later by Shiftas, or Somali bandits. The pretty little hotel has been deserted since then. I had no idea that the area was so dangerous when I slept there.) We bought a pile of old whodunits and packed up a whole carton of canned food. These items we carefully wrapped and lashed together with a long piece of cord to which we attached an empty drum, the two holes in which had been plugged and sealed. Once more we flew northwards along the lake-shore. We swooped low over the little camp and its three lonely occupants and dropped our oil-paper packets in the shallows near the shore. The idea worked, so the three men now had enough supplies for several days.

This rescue operation was the sole reason for everything that happened to us in the next few days.

Not that we suspected anything at the time. We flew diagonally across the lake, heading more or less due west towards the extreme northern tip of Uganda, which is enclosed by Kenya and the Sudan. It was late afternoon, and the area was virtually uninhabited. The map resting on our pilot's knees showed nothing but a few lines and patches of shading to indicate the rough position of rivers and mountain ranges. The greener the land below us, the cloudier the

We must land before dark

skies above. It was a landscape totally unlike our own, where every couple of miles brings another village, town, or high-road, where villages cluster together with increasing density and where no valley is without at least a couple of houses. We had the earth and sky to ourselves.

The sky grew darker and darker as we flew on through sheets of rain, the storm-clouds above us rent by lightning. Six o'clock came, then six-thirty. Nightfall was only half-an-hour away. Flying our light aircraft in darkness would mean certain death. There were no illuminated runways here, no airports, no blind landings controlled by radio. We would inevitably crash into a hill or a tree.

But where was Kidepo, Uganda's third and latest national park? It had only just been established, so there were few visitors and no proper roads, and the game wardens' quarters probably looked little different from the scattered huts which we occasionally glimpsed in the mountains. It also turned out that our pilot had only been in East Africa for three months and was unfamiliar with the area.

We flew lower and lower, following the only car-tracks which might have been described as a road, but they led on and on into the green distance past native huts, and the rain in front of us was intense. A menacing bank of jet-black cloud loomed up between the mountains, so we turned and flew back. A few buildings with corrugated-iron roofs hove into view – possibly the park headquarters. Whatever they were, we had to get our wheels down as soon as possible. Not far from the building was a level strip of ground which gave promise of a smooth landing. We buckled our seat-belts and swooped down. Darkness was already falling.

We climbed out. It was good to feel solid ground beneath our feet again. The buildings were much farther away than they seemed from the air, but two trucks had already emerged and were jolting towards us. Two more approached from the other side. It was clear that we had happened on what was, by local standards, a hive of civilization.

Then the trucks halted a hundred yards from us and forty or

Greeted by machine-guns

fifty black soldiers in steel helmets climbed out. They set up machine-guns and trained them menacingly on our plane. A dark-skinned Arab, presumably their officer, approached us brandishing a pistol. The soldiers bore down on us from all sides, rifles at the ready and safety-catches off. The officer knew no English, but two of his men did. We had inadvertently crossed the Ugandan frontier and flown some distance into the Sudan. Our baggage was unloaded. The officers, two of them now, displayed mounting excitement when five cameras, various lenses and a movie camera came to light. They obviously thought we were spies. Our pilot turned out to be carrying a pistol, which was news to me.

We had landed in a police-cum-army camp. While we were being interrogated by candlelight in the guard-house, one of the English-speaking soldiers assured Alan Root that we would be shot. The whole thing struck us as ludicrous and absurd.

We were billeted in a room belonging to three N.C.O.s, who hospitably cleared out their clothes and put their beds at our disposal. One of them even lent us a small radio. I switched it on and learned that a major insurrection was raging in Equatoria, the Sudanese province whose borders we had violated. Fifteen hundred Africans had been shot, together with a couple of Englishmen reputed to have been spies.

I began to feel a trifle uneasy. Nobody had any idea that we were in the Sudan. If we did not reappear it would be assumed that we had crashed somewhere in the wilderness. If we were not back in Entebbe inside three days the authorities might send out a number of aircraft to look for us, and we should have to foot the bill later. Rescue operations come expensive. I drafted some telegrams. The chief of police promised to transmit them by radio immediately. Sleep was quite impossible. All the films we had taken at Lake Rudolf and elsewhere were confiscated. Naturally, no one believed us when we pointed out that the light would have made it impossible for us to photograph 'military objectives' in the Sudan so late in the evening.

Nobody could say what would happen to us. The local commander had reported our detention to Torit, the district capital, and was awaiting instructions.

Whenever we went to relieve ourselves next day, two soldiers armed with rifles escorted us. I experimented to see if I could take a different route. The soldiers raised no objection, so I walked to the neighbouring village under military escort. The mission church was barred and deserted, but the local inhabitants greeted me in friendly fashion. They opened up the hospital dating from British times and showed me the modest surgery. There were few drugs in evidence, but the African medical orderly seemed to be doing his best. The large open court-house in the middle of the village still bore a sign inscribed in English and Arabic with the words: 'It is forbidden to urinate in court'. The black village headman, who spoke good English, came up and introduced his wife and children. I distributed visiting-cards to various Sudanese, determined to leave at least some token of our presence. The school-house, which the villagers gladly opened up for my inspection, was empty. All schools were temporarily closed, they informed me. Since when? Since two years ago.

Interspersed with the villagers who gathered round us were a number of entirely naked young men. No one seemed to find this remarkable. The scene of our arrest was apparently called Ikitos. I remembered that my friend John Owen, Director of the National Parks of Tanzania and now based at Arusha, had been a district commissioner in the Southern Sudan during the British administration. Sure enough, his name rang a bell. The headman asked me if John Owen had grown very old in the meantime and what his daughters were doing. One of them had been born there and named after a village in the neighbourhood. The inhabitants of this remote and isolated corner of the world were obviously glad to hear something of what went on outside, but their nervous and extremely guarded manner told one that they were downtrodden. John Owen had actually been the last British district commissioner of this particular area. He was based at Torit and often presided over the local court.

No possible escape

I drafted more telegrams, wondering if they would ever reach their destination. Lying sprawled on our beds, we read all the books and periodicals we had with us. You see, I told myself, at last you're getting the enforced holiday all your friends have been prescribing for so long.

The Arab officers and N.C.O.s treated us with great cordiality. They brought us cheap canned meat, a carafe of drinking-water, tea, sugar, and a packet of soggy and very ancient wafer biscuits. We received a personal visit from the commanding officer, who did not speak English, this time attired in a long Arab robe. It even appeared that a goat had been slaughtered for our benefit, because we were brought an entire bowlful of roast meat for supper. The local inhabitants were spoken of with great condescension and a certain measure of contempt. The troops armed with rifles and pistols were there for our protection, we were told, not to guard us like prisoners, because the 'natives' were very dangerous. The atmosphere was all very Arab and amiable.

Each of us had a book to read, but three books would not go far among three men who had nothing to do but lie on their beds from dawn till dusk. We were feeling filthy, too, because our main baggage was still at Entebbe airport. We made plans to sneak into our plane, open the throttle and hedge-hop back across the border, or at least establish radio contact with the big airliners which must be winging their way through the sky high above us. Then we thought of the heavily armed troops who were camped round our machine even at night. By the third day we were planning to break out and run for it – an idiotic idea, of course, because it would have entailed many days' march through unknown territory. Finally the C.O. came to inform us that we were to be conveyed to Torit, the district capital, and from there to the provincial capital of Juba, on the Nile.

Five trucks were loaded with fifty or sixty steel-helmeted, heavily armed soldiers equipped with machine-guns. Each of us was put in a different vehicle. I was seated beside the C.O., who could only tell me the names of villages as we passed them. These were few and far between. Picturesquely situated among wild and rocky mountains, they consisted of circular huts interspersed with little stilted store-houses, each settlement being surrounded by a tall jagged stockade. The lengths to which our captors were going on our behalf made us feel very grand. The roads were incredibly poor and worn, and quite deserted. There was not a car or pedestrian to be seen. The African villagers peered at us timidly from their huts.

The commanding officer, a handsome young man with black skin and finely chiselled Arab features, explained to me in sign-language that his back was hurting him. The hours of bumping and jolting must have meant agony to him. He glanced at his watch and brought the entire convoy to a halt. Silently, he removed a small rug from the truck and spread it in the middle of the road. Then, kneeling down, he raised his arms in the direction of Mecca and pressed his forehead to the ground. He was the only one who prayed. The rest stood and waited.

The country-side reminded me of Guinea in West Africa. It must have been a wonderful place in more peaceful times, and I quite understood why John Owen spoke so enthusiastically of his years there. Just as they used to in the West African interior fifteen years before, the jet-black girls of the district went about gracefully erect with their round breasts serenely exposed.

A reedbuck darted across the road and halted in the scrub a hundred yards away. The commanding officer grabbed his rifle and leapt out of the truck. He fired but missed, much to my secret delight. This antelope was the only wild animal I saw on the trip. Few antelopes or elephants survive in areas where so many people carry guns.

At last we saw some vehicles heading towards us – the only ones we encountered in 140 miles. It was an escort: an armoured car in front, then trucks full of heavily armed soldiers, then lorries packed with men and supplies, and finally more soldiers. The situation was obviously tense and the local African population in a state of turmoil. It was a perfect setting for an armed insurrection. If someone

A hotel at last

had fired at us from the surrounding rocks and clumps of bushes the Arab soldiers would, at most, have taken cover behind their trucks. They certainly would not have ventured far into such wild and hazardous terrain. Our escort was more than a mark of courtesy – it was there to protect us rather than keep us prisoner.

We were to spend the night at Torit before setting off next morning with another escort for Juba, the provincial capital. I sat in Owen's old office, where it was obvious that nothing had changed over the years. I insisted on speaking to the general and actually persuaded him to give orders that we should be escorted to Juba by the same soldiers. Earlier, we had dined in the officers' mess, plunging our fingers Arab-fashion into the bowls of roast meat, rice and vegetables and dipping each mouthful in salt or pepper according to taste. There was only water to drink.

Late that night we crossed the Nile in one of the large and decrepit car-ferries. A searchlight illuminated the vehicles as they drove down the rutted incline of the river-bank and on to the steamer by way of two wooden planks. When we arrived I asked the commanding officer in a casual tone if he had booked rooms for us at the Juba Hotel because I had no wish to spend another night in police custody. He looked surprised but escorted us from the police station to the hotel by car.

I had not been there since Christmas 1957, when I and my son flew up the Nile from the Mediterranean and on to Lake Victoria and Serengeti. The manager dutifully asked to see our passports, which were no longer in our possession. Everything was as it had been – the big ceiling-fans, the worn and obsolete oilcloth-covered chairs, the sand-strewn garden paths leading from the restaurant to the wing containing the guest-rooms.

The only difference was, there were no Europeans to be seen apart from two British pilots drinking beer in one corner. It was not true that two Britons had been shot, they told us. They had simply been arrested for photographing dead bodies and blazing buildings during the rebellion. Apparently, they were employees of a tobacco company. After we had eaten I received a visit from the police chief of Equatoria, a dark-skinned Arab who had done his training in Germany. He was accompanied by his wife, a German girl from Wiesbaden, who was carrying their first child in her arms. Life in this southernmost Sudanese province obviously agreed with her, even though the area was strictly sealed off. She was hoping to pay her first visit to Germany in a few months' time. The police chief courteously proposed that we should travel to Khartoum next morning by freight-plane, a five- or six-hour flight which took in numerous detours and intermediate stops. Meanwhile, our British pilot had chummed up with his two compatriots, who were employed by the Sudanese army and were scheduled to fly a military aircraft direct to the capital early next morning. We persuaded the police chief to send the three of us to Khartoum in their brand-new plane, escorted by the local commander from Ikitos. Flying-time would be only three hours.

At the airport a party of bearded Africans in white gym-shoes crowded round the aircraft. They were Congolese rebels who had been driven out by Tshombe. The British pilots refused to take them along without a permit, which was a pity. I should have enjoyed questioning them during the flight. It had been alleged by European newspapers that the Sudan was allowing arms from China to filter through to the Congolese rebels. Whatever the truth, arms from the Congo were certainly being smuggled across the Sudanese frontier for use against the Arab authorities. It is easy to distribute arms in Africa but hard to call them in again. Millions of Africans in the Southern Sudan were cruelly exploited for centuries by Arabs from Egypt and Khartoum, who sold them on the international slave market. The Africans had bettered their lot under British rule but now felt that they were back in the clutches of the old slave-traders. Some of them had become Christians and many had attended school and enjoyed the benefits of orderly government – hence the uprisings. The British should, before their departure, have guaranteed at least a measure of local self-government in the new and independent Sudan.

As we flew high above the clouds, our pilots conversed by radio

My cables not sent

with air traffic control in Nairobi, the Kenyan capital. We learnt that a search had already been in progress for days. It was feared that we had come to grief because the battery of our plane had been empty the night before we took off from Lake Rudolf. Now, at last, people in East Africa knew where we were. I felt much easier in my mind.

At Khartoum we were passed from one department to another. Police officers from the Ministry of the Interior interrogated us in ascending order of rank, plying us with coffee and cigarettes the while. Once the police files on us were complete, the Attorney-General would decide if we were to be brought to trial for unauthorized entry and suspected espionage in a combat zone. I protested against our detention and pointed out that there is an obligation under international law to assist aircraft that have made forced landings.

'Yes, but you came down in a restricted area where fighting was in progress.' I retorted that the Sudanese Government had not announced any such restrictions on East African air traffic, presumably because they wanted to keep the rebellion dark. I asked if my telegrams had been dispatched and was shown them neatly tucked away in the files. Then, but only then, did an official pass them to the cable office by telephone in my presence.

That evening, in the ultra-modern but almost deserted Sudan Hotel, I read in a Khartoum newspaper that 'two German professors' had been detained in Equatoria on suspicion of espionage, namely, 'Prof. B. Hard and Prof. Griizinik'. The editorial staff had made two separate people out of Bernhard Grzimek. I had long been aware that the Sudan, being an Arab country, did not boast a German embassy, but the German Consul-General, Dr Theodor Mez, who is now attached to the French embassy, made feverish efforts on our behalf. He signed a bail form which enabled the three of us and our British companion to be set at liberty and accommodated in the hotel. Dr Mez spoke fluent Arabic.

If it came to legal proceedings I should have plenty of time to sit on the hotel terrace and watch the brown-sailed Nile barges glide

I keep quiet about my books

past. I racked my brains for a way of convincing the Sudanese that I was only interested in wild animals, not politics. Then another guest at the hotel told me that English editions of my books were on display at an Arab bookshop. We drove there to find two ladies leafing through them and discussing reports that the man who had written them was languishing in a Khartoum gaol. The Arab bookseller readily lent me the books so that I could parade them at the Ministry of the Interior. He even refused to take a deposit.

Very close to our hotel lay Khartoum Zoo, which contained a Frankfurt-born hippo and was the source of our Frankfurt shoebills. I had known the director for many years, also the director of the Sudanese national parks, whose office was at the zoo. He had visited me in Frankfurt and his dark-skinned son spoke fluent German. But, although he invited us to a non-alcoholic supper in his blossom-filled garden, he could hardly intervene in police or legal proceedings. And so we continued to telephone, telegraph, and pull strings. Khartoum was a delightful place, but I had no wish to stay there for weeks on end, not even at government expense.

During one of the warm nights that followed I was struck by an unpleasant thought. I recalled that in my book about Serengeti I had described how we came across a concentration camp, complete with watch-towers and machine-gun nests, near Juba. A European doctor employed by the Sudanese government informed us that the Arabs occasionally shot twenty or thirty of the natives imprisoned there, and I had made some rather unfriendly comments about this.

'Is that in the English edition too?' I asked Alan. He nodded. Too bad. If one of the Sudanese officials found out, the Attorney-General would be even less likely to believe that I had strayed into the forbidden province without an ulterior motive. I decided not to show the Attorney-General a copy of *Serengeti shall not die* after all.

There is a special technique for dealing with bureaucrats. The young British consular official laboriously showed the porter his credentials and requested an interview with the Attorney-General.

President Nyerere intervenes

We simply announced that we had an appointment with him and marched past the desk at a brisk and purposeful pace. More coffee-drinking and courtesy, but the police had not yet forwarded our files. The Attorney-General was still unfamiliar with our case and could take no decision.

'What is the penalty for unauthorized entry?' I demanded. I couldn't hang around Khartoum for weeks until the case was settled. What about the possibility of bail?

Time was no object. However, the senior civil servant actually telephoned the police and asked for our files. We descended the stairs to find the consular official still negotiating desperately with the porter.

That afternoon Dr Mez rang me in a state of exaltation. A few hours earlier the Sudanese prime minister had received a cable from Dr Julius Nyerere, President of Tanzania. I was, after all, an honorary member of the Tanzanian civil service. Cables had also been sent by U.N.O. and the president of the German Bundestag, Dr Gerstenmaier. The Sudanese cabinet had met on our account and decided to release us forthwith.

I felt extremely important – in fact it would not have surprised me to hear that my case had been raised at the Afro-Asian Conference. The Minister of the Interior expressed his regrets. There was no further insistence on developing our precious films at Khartoum, and our confiscated equipment was returned with enormous ceremony. I was appalled to find that our cameras had made the bone-shaking 140-mile journey by truck piled on top of each other in an unpadded cardboard box, but everything seemed to be intact. Having lost a week, I could not afford the time to return to Kampala. Shabby and dirty, I grabbed my toothbrush and photographic equipment and boarded a jet for Rome, where I hurriedly switched to a Japanese airliner. I arrived in Frankfurt unshaven. My house-key, car-keys and cheque-book were still at Entebbe on Lake Victoria.

13. The much-maligned gorilla

Philosophy resembles a tidy apartment to which the animal is not admitted.

ALBERT SCHWEITZER

For a hundred years, from its discovery in 1860 until today, the gorilla has been regarded as a huge and devilish jungle monster which furiously attacks and destroys human beings. The zoologist R. Owen was one of the first men to give an account of these anthropoid apes, which range in colour from dark brown to blue-black. He alleged that natives creeping furtively through the gloom of the tropical rain-forest were sometimes alerted to the presence of one of these fearsome giant apes by the sudden disappearance of one of their companions. The unfortunate man would be pulled up into a tree with no chance to utter more than a brief cry of alarm. A few minutes later he would fall to the ground again, a strangled corpse. Only a few years ago a German sporting journalist stated that he had been compelled to 'execute' a male gorilla in West Africa because it had assaulted every female gorilla it could lay hands on, broken into native huts, killed a mother-gorilla and stolen her young. I was able to prove subsequently that this 'murderous' gorilla was, in fact, a female. All in all, precious little was known until very recent years about the disposition and living habits of these massive creatures which are so closely related to us.

Professor George B. Schaller of the Serengeti Research Institute spent twenty months living among mountain gorillas and observing them every day. He submitted a 430-page scientific report (G. B. Schaller, *The Mountain Gorilla* Chicago 1963) which is probably the most comprehensive work on an African animal that has ever been written. With Schaller's permission, I sent my associate Alan Root

G. B. Schaller's gorillas

and his young wife Joan to film and photograph gorillas in that part of the Congolese Republic where Schaller himself had struck up a sort of friendship with them. The Roots spent from September to December 1963 in the area.

Gorillas, especially those of the mountain variety, are extremely difficult to film. They are timid creatures which spend much of their time screened by undergrowth and vegetation, quite apart from being black in colour. The photographer seldom has enough light, even when using highly sensitive film. The few shots of free-ranging gorillas that have so far figured in two or three commercial films are really of animals previously encircled or captured with the aid of hundreds of beaters and released in open-air enclosures for the camera-man's benefit. We were afforded a good opportunity of filming mountain gorillas leading a natural and unconstrained life in the Virunga Range because uninterrupted coexistence with Schaller had accustomed them to human beings. Unfortunately, poaching had since become rife in the area, so I realized that it might only be a few months before these few hundred mountain gorillas regained their fear and timidity in the presence of man.

The mountain range in question consists of six extinct volcanoes which rise from one of the most fertile and densely populated regions in Africa. It was primarily to protect the mountain gorillas which inhabit these volcanoes that the Belgians established the Parc National Albert in 1925. Covering an area of 1,200 square miles, or slightly larger than the East Riding of Yorkshire, it also includes broad plains and lakelands where no gorillas dwell. The six volcanoes sit right on the border between the Congolese Republic and the new state of Ruanda, both of which used until recently to be governed by Belgium. Although the new Congolese administration began, despite the civil war, by expressly protecting mountain gorillas, they are constantly menaced by farmers who want to fell more forest and put more land under the plough, by hunters who shoot them for meat, and by Watusi herdsmen who encroach on their territory with huge and uneconomic herds of cattle.

Gorilla population overestimated

The area in which these grave-faced anthropoid apes dwell, and in which scientists and observers must seek them out, hardly matches the popular conception of a tropical paradise. It lies at altitudes of between 8,500 and 14,000 feet and has a heavy rainfall. The mountain forests are often pervaded by cloud and mist, temperatures can plummet to below freezing point, and hail-stones the size of marbles are not uncommon.

It seems likely that the heart of Africa – which is the only place where they occur – currently supports a total of five thousand mountain gorillas, probably no fewer but certainly no more. They are not averse to frequenting places where man has been at work, in other words, the vicinity of plantations and villages, roads and mines. Wherever large trees have been felled, gorillas can obtain more food in the shape of plants, bushes and saplings, not to mention the products of human agriculture. It is eighty years before real forest trees take root of their own accord, gain a hold and attain a height at which they become indistinguishable from the original forest. The animals have always been assumed to be as numerous in the boundless forests as in the neighbourhood of roads and settlements, which has given rise to grossly exaggerated estimates of their total population.

The Schallers and the Roots lived in a log cabin and in tents, principally in the Kabara district of the Virunga Range. Of the four or five hundred mountain gorillas inhabiting the slopes of these volcanoes, about two hundred lived there in ten separate families or groups. George Schaller spent 466 hours observing them at close range and recorded 314 encounters with them. By the time he left, six of the ten groups had become wholly accustomed to him.

Undergrowth and bushes make it far from easy to watch gorillas going about their daily business. The advantage of the Kabara district was that few if any of the gorillas there had ever encountered man. They seldom fled when approached by a single observer. The group-leader, a huge and massive male, would turn to face the new-comer and utter a few cries, whereupon the females and

Mock attacks

young gathered near him and observed the unfamiliar creature with equal attention, often from raised branches or tree-stumps. They generally resumed feeding or resting after a short time. If the observer busied himself with an object which was strange to them – a movie camera, for instance – they craned their necks or moved to a vantage-point which afforded a better view.

Groups varied between 2 and 30 head, the average at Kabara being 17. It was inadvisable to walk straight up to the gorillas, make abrupt movements, break into sudden speech or stare back at them, or they would take fright. No genuine attacks were recorded.

Sometimes, for no apparent reason, curiosity overcame the gorillas. One day, a group which Schaller had already encountered and observed 76 times gradually edged closer to a distance of 10 yards and scrutinized him with extreme curiosity. A female with a three-months-old baby at her breast edged still closer, reached up and gave the branch on which Schaller was seated a sharp push. Then she looked up as though to see what his reaction would be. Shortly afterwards a half-grown male followed suit, and one lone

Page 213:
Gravely and meditatively, the huge manlike creatures gaze across at the photographer. They only take flight if someone stares at them too fixedly or ventures too close. The careful observer can sometimes remain within twenty or thirty yards of a group of gorillas for hours on end.

Page 214, above:
Prince Bernhard of the Netherlands was the first person to wire his congratulations on the birth of the gorilla twins Alice and Ellen on May 3rd 1967. Although twins, they differ considerably in appearance and even more so in temperament.

Page 214, below:
A baby gorilla which has been rejected by its mother makes as much work as a human baby, if not more. It does, however, become ardently attached to its human foster-mother and generally retains a lifelong preference for human beings as opposed to other gorillas. This is Max, the first gorilla to be born in Germany.

Page 215:
The female gorilla Toto attained an age of at least 36 in captivity, and is the second most long-lived gorilla on record. She adored cats and became such friends with two that she carried them around everywhere and the cats repeatedly visited her quarters of their own free will. Toto, who had lived among human beings from babyhood onwards, proved incapable of making friends with a male gorilla in later life.

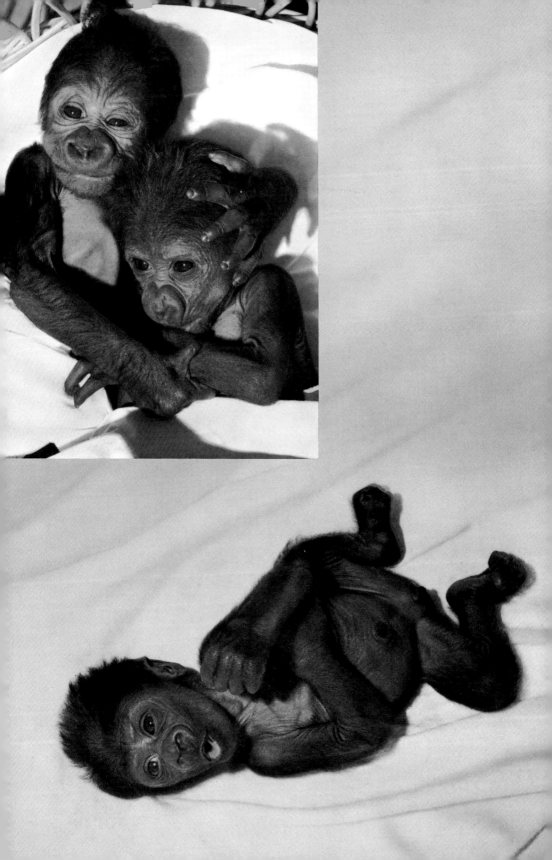

No sexual rivalry

female actually climbed Schaller's branch and remained there for a few seconds. More often, mock attacks were made by the males, foremost among them the group-leader, presumably to test Schaller's reactions or drive him away. One large male dashed to within 7 yards of him in this way but eventually settled 25 yards away.

Seldom has any creature on earth been so disgracefully misjudged as the gorilla. To anyone reading the entries which Schaller made in his voluminous diaries during long hours of observation, this is the only possible conclusion.

Most creatures, man included, take possession of specific territories and defend them fiercely – indeed human beings often kill each other on that account. The gorilla group also has a preserve of its own, a territory measuring 10 to 15 square miles. The animals, which are almost entirely terrestrial, spend their whole time ranging the area in quest of food. The distance covered in a single day may vary from only a few hundred yards to three miles. Other groups of gorillas inhabit the same area, but there is no fighting when two groups meet. The group-leaders exchange stares and sometimes, though not always – threaten one another, but members of two groups may even join forces and then disperse again. Schaller never observed any quarrelling, let alone actual fighting, among gorillas.

Fighting over females is equally rare, and sexual activity in general plays a far more subordinate role among gorillas than it does among most other apes. The group-leader will permit another

Page 216:
The mountain gorilla inhabits the slopes of the extinct Virunga volcanoes, which overlook one of the most densely populated regions in Africa. Situated close at hand in the Congo are other volcanoes, some of them active, like the Nirangongo whose lava-filled crater is illustrated here. It rises to a height of 11,384 feet.

Gorillas spend most of the day on the ground, not in trees. They have no natural enemies save man, which is why they take relatively little trouble to move silently. They feed almost exclusively on leaves, together with the pith and bark of some hundred different varieties of plants; also – to a lesser extent – on fruit. They never touch food of animal origin.

Pregnant gorillas

male gorilla to flirt and even copulate with one of the females in his group only 10 feet away. Pregnancy is almost imperceptible in female gorillas because of their extremely powerful stomach muscles. To judge by the births that have so far occurred in zoos, the gestation period is just on $8\frac{1}{2}$ months as compared with just under 8 months in the chimpanzee and 8 months in the orang-outang. Many female gorillas develop temporarily swollen ankles during pregnancy. Parturition takes only a few minutes. After giving birth in a reclining position, females sever the umbilical cord and carry their young by clasping them to their breast. Unlike many lower species of apes, young gorillas cannot cling to their mothers unaided. By the beginning of 1969, 21 gorillas had been born in zoos under human supervision. There were also 6 miscarriages, but 18 of the 21 mature babies survived. 8 of these births were divided equally between the Basle and Frankfurt Zoos. Virtually all zoo-born young are rejected by their mothers and have to be artificially reared by human beings. This is probably because the mothers, like our Makula at Frankfurt, also entered human custody at an early age and never learnt how to handle baby gorillas in the wild. 5 young gorillas have been reared from babyhood in my own home, as I have described in my book *Twenty animals, one man*. Our gorillas at Frankfurt Zoo behave in exactly the same manner. Young are born throughout the year, so there is no preferred season for breeding.

Our female gorilla Makula, now adult and living in Frankfurt Zoo, was herself such a tiny baby when she came to us that she could not even raise her head. We had to keep her in a cot beside the bed at home, like a human baby. Baby gorillas develop about twice as fast as young children.

Our first Frankfurt-born gorilla, Max, who came into the world on June 22nd 1965, weighed 4 lb. 10 oz. at birth and 36 lb. after twelve months. He acquired his first milk-incisors during the sixth week of life and his first molars during the seventeenth. He could follow moving objects with his eyes by the second week, laughed when tickled, and raised his head from a prone position. When he was ten weeks old he could distinguish his foster-parents from other people and roll from his tummy on to his back. At nineteen weeks he was walking on all fours, at twenty-six weeks he could rise on his hind legs and was drumming on walls, and at thirty-four weeks he could already take a few steps in an erect position.

On May 3rd 1967 the same female produced female twins. They were uniovular, however, and differ considerably in temperament and weight.

In the Kabara district, young gorillas join their elders in feeding on plants at two-and-a-half months. It is clear that their diet consists mainly of plants by the sixth or seventh month, but many continue to supplement it with mother's milk until they are eighteen months old. They can climb at six or seven months.

A female gorilla with young of her own is not unfriendly towards other babies. She will not push them away even if, as sometimes happens, they climb on her lap and suck alongside her own. Massive and powerful though they are, male gorillas patiently allow young to play round them or climb on them. One female gorilla in the Virunga Range area had an eight-month-old baby with a big wound in its rump. Unlike the rest, it never rode on her back because it was manifestly too weak to hold on. The mother always carried it carefully on her arm so that no part of the wound was in contact with her body – in other words, she held it belly-downwards. She used to examine the cut closely from time to time and pick at it briefly with her fingers. On one occasion another female who had lost her own baby two months earlier bent over the injured youngster and brushed its face with her lips.

Very young babies are often clasped in both arms when the mother is seated. One female carried her dead baby around with her for four days. No instance of thumb-sucking was observed. The young remain with their mother and sleep in her nest until the age of three. The bond is not dissolved until they are four or five, but not in such a manner that the mother brusquely chases her offspring away. A growing youngster often continues to sleep with its mother after she has had another baby and still grips the

A white gorilla with blue eyes

fur on her rump while walking. Mothers are sometimes quite gentle in their efforts to detach a half-grown youngster's hand. Some gorillas tend to have brownish fur in their early days, particularly on the head, and many retain this brownish light on the crown of the head throughout their lives. In 1966, a male gorilla aged about two was found in the Nko Forest in Angola. It was a perfect albino, pure white with blue eyes. The animal has since lived at Barcelona Zoo.

Male gorillas become sexually mature between the ages of nine and ten, females at six to seven. Little can be said about their lifespan in the wild, but they have survived for as long as 38 years in zoos. A full-grown male gorilla is an impressive figure, and weighs between 300 and 440 lb. Females weigh between 150 and 240 lb. The big male known as Phil in St Louis tipped the scales at 770 lb. when he died but was probably overweight. At the age of ten, males acquire a silver-grey back. Only these silver-backed males can become 'headmen' or leaders of family groups, but even without their silver saddle one could hardly mistake their gigantic size and majestic appearance. In contrast to baboons or primitive human beings, they exercise their supremacy without strife, blows or biting. There is no squabbling over food. The fact that gorillas have a certain ranking order becomes most readily apparent when they meet or overtake others of their kind on a narrow path, in which case the lower-ranking animal has to stand aside. If it fails to do so voluntarily, a gentle tap with the knuckle or hand generally proves a sufficient reminder. More rarely, the senior animal thrusts its inferior away with both hands or barges it aside by calmly continuing its advance. The group-leader makes no attempt to oust adult but inferior males from the group or keep them at a distance. Equally little effort is made to prevent male outsiders which inhabit the forest singly or in pairs from joining the group. Adult males can be seen sitting only 5 or 6 feet apart.

The young are attracted by the group-leader. As soon as the group calls a halt, one or other of the youngsters leaves its mother

No mutual delousing

and sits down beside him or romps about on him. One group-leader was seen with four youthful admirers. When one of the babies smacked him in the face the massive male simply turned his head away. He even allowed them to grab him by the fur, ride on his back for a spell, or climb on his lap when he was seated. If they became too much of a nuisance, one glare was enough to quell them.

Everyone knows that the practice of mutual 'delousing', to use the popular phrase, is widespread among monkeys. In reality, this has nothing to do with the catching of lice because monkeys are generally free from lice and fleas. It is really a form of reciprocal grooming designed to express solidarity and affection, and more or less corresponds to an exchange of caresses between human beings. Rarely indulged in by gorillas, it can be observed most often between mothers and young, sometimes between powerful silver-backed males, but never between the younger black-furred males. Many young baboons and chimpanzees delight in being 'deloused' by their mothers and positively clamour for more, whereas young gorillas often seem to find it irksome and resist like children who dislike being washed.

Chimpanzees and many other monkeys are playful creatures. Young gorillas sometimes wander away from their mothers and play near-by when the other members of the group have lain down to rest. The adults do not join in. In free-ranging gorillas, playfulness ceases at about six or with the onset of sexual maturity. Games favoured by the young include tag or defending a tree-stump or hummock against attack by other youngsters – much the same activities, in fact, as those performed under various names by young human beings. Other pastimes indulged in by young gorillas include climbing, swinging from branches, jumping, sliding, flattening plants, turning somersaults, and racing around in a 'giddy' way.

The whole group takes its cue from the silver-backed leader even though he does not appear to give orders of any kind. If he moves off after a rest, all the others follow suit without maintaining any form of order. He is the most alert and excitable of all, screaming and uttering threats long after the others have calmed down. He is also the most nervous, and likes to conceal himself in foliage. Sometimes a group contains two silver-backed males of which the junior maintains a certain distance between himself and the leader. It may also comprise one or two black adult males, six or seven females, two or three immatures, and four or five young. Gorillas yawn, pant, cough, belch, hiccup and scratch like human beings. They communicate by emitting 22 distinct sounds but possess no form of speech, this being confined to the human race. Mutually tolerant though gorillas are, the pronounced capacity for friendship found among chimpanzees does not appear to exist in gorillas of the same group.

The silver-backed male quells members of his group, as well as animal intruders and human beings, by resorting to strange and awesome intimidation behaviour reminiscent of a dance of rage. He starts by uttering a series of screams, sometimes breaking off to take a leaf between his lips. After that the screams succeed one another in even quicker succession and he rises on his hind legs to hurl leaves and twigs in the air. He beats his chest several times, then scuttles sideways on all fours, tearing up plants in the process. Finally, he drums on the ground with the flat of his hand.

Youngsters four months of age can also be seen beating their chests and tearing up plants, but only adult males utter screams during this form of intimidation procedure. Male dances performed by American Indians, Africans and other primitive peoples serve the same purpose, just as parades, protest marches and manœuvres do with us. The object is to make an impression and intimidate without resorting to actual violence, let alone killing.

Gorillas which really intend to attack behave quite differently – once again, not so very unlike human beings. When a gorilla stares keenly into the eyes of another gorilla or an approaching man, this is meant as a threat and construed as such. For this reason, observers and camera-men must avoid looking at gorillas too closely. The menacing male sometimes knits his brow and compresses his lips as well, emitting short staccato grunts. Finally, he jerks his head – often, too, his body – in the enemy's direction as if about to charge. In the last resort he really does charge, but almost invariably halts a few yards short or swerves and runs past. The hunter who happens to have a rifle with him always shoots the animal 'in self-defence'. There have been many reports of such killings in the past few decades.

It can, of course, be dangerous to lose one's nerve at the giant creature's approach and run away. The gorilla will generally pursue a fugitive on all fours, bite him in the legs, buttock, arms or back and then lumber off. Very seldom will it linger over him or inflict further injury, far less kill him.

Human beings also interpret an exchange of stares as provocation. Only a few decades ago, members of German students' clubs used to take umbrage when stared at fixedly by someone in a restaurant. They invited the offender outside into a corridor or wash-room, swapped slaps in the face and calling-cards, exchanged seconds, and later fought a duel. Thus, human beings still have in them a great deal of inherited behaviour which they share with the large anthropoid apes. Although student brawls and duels resulted in very few fatalities, the death-rate among gorillas is probably far lower.

Gorillas have a built-in mechanism which inhibits mutual injury or slaughter. The female or subordinate male averts the eyes or, more usually, the entire head when glared at by a superior, merely as an indication that there is no desire to join battle. The male gorilla at the Columbus Zoo averted his gaze from the female whenever he felt amorous, so as not to intimidate her. A pair of adult gorillas at New York, when finally put together after a long period of acquaintanceship through a barred partition, also tended to avert their eyes and heads when approaching one another. One day the female lashed out at the male and tore a long gash in her forearm because she had caught it on one of his canines. She busied herself with the cut for an entire day. The male gorilla was noticeably affected by this incident. He left the female severely

Walk, never run!

alone for some time afterwards and always retreated from any part of their living quarters soon after she entered.

A gorilla which is stared at or approached by another gorilla or a human being has three ways of conveying friendly intent. The first is to look away and turn the head to one side. Secondly, the head may be shaken back and forth and, finally, the animal may lie flat on the ground, face downwards. Schaller tells us that he several times conducted simple experiments to ascertain the occasions on which a gorilla might be likely to incline its head. One silver-backed male sat ten yards away, watching him. When Schaller focused his gaze on the animal's face it became perceptibly unsure of itself and averted its head. Schaller continued to stare at it, whereupon it started to move its head from side to side. Eventually it stood up, beat its chest, and moved off. A black-backed male regarded him in similar fashion, also from a distance of ten yards. When Schaller nodded, the animal averted its gaze. When Schaller, in his turn, stared at the animal keenly, it started to nod its head. They kept this up for about ten minutes. Black-backed males sometimes nodded their heads when they passed very close to him, and he employed the same movement when he unexpectedly came upon them a short distance away. The gesture, which seemed to soothe them, apparently signified an absence of hostile intent.

A gorilla which wants to avoid being 'out-intimidated' by another male or display its subordination very clearly as soon as a higher-ranking animal enters the vicinity performs a sort of obeisance. This entails lying face downwards with head pressed to the ground and arms and legs drawn in beneath the body so that only the broad back is exposed. A young gorilla sometimes places one hand on the back of its head as well. This not only protects the most vulnerable parts of the body but also, and more especially, inhibits all aggressive urges in the senior animal. We human beings behave in a startlingly similar fashion. After all, the man's low bow and the woman's curtsy are nothing more nor less than tokens of respect and submission. When Man Friday first encountered Robinson Crusoe he not only prostrated himself but actually placed Crusoe's foot on his neck. Many primitive peoples still approach their rulers on all fours or sprawl on their stomachs as we ourselves used to do in less recent times. Among the Nyakyusa of South-West Tanzania, a woman who greets a man has to bow and avert her gaze. Bowing the head as a form of greeting probably stems from the same wish to demonstrate friendly intentions as that which motivates the gorilla. It may well be an inherited or innate mode of behaviour.

Dr Schmitt, one of my associates at Frankfurt Zoo, has in recent years performed electrophoretic and other very exhaustive blood-tests on anthropoid apes, human beings, and other monkeys. On this basis, too, gorillas are very closely related to us, less so than the chimpanzee but more so than the Asiatic orang-outang. The temperament and living habits of the gorilla seem to bear this out. In general, though, their behaviour makes a far more attractive impression than that of chimpanzees, which seems to be an unpleasant reflection of our own worst human characteristics.

Chimpanzees are far more quarrelsome. According to research carried out by Jane Goodall in Tanzania's small Gombe National Park, chimpanzees living in areas of semi-steppe habitually kill and devour smaller monkeys. Where free-ranging gorillas are concerned, no one has ever found evidence of carnivorous habits. Schaller inspected several thousand heaps of dung left by wild gorillas without finding a single trace of hair, chitinous insect casing, bone, skin, or other food of animal origin. Gorillas quite readily eat meat in captivity, but this is probably attributable to the complete change of diet and lack of protein.

In the wild, they took no notice of freshly killed animals and passed them by. One group of gorillas rested within 10 feet of a brooding dove. Although they could not have failed to see the bird, they did not molest it. Gorillas live mainly on leaves, shoots, and the pith of plants and saplings. They also tear off bark and eat it. The animals almost invariably use their hands when eating and seldom bite off vegetation with their teeth. Diet is another point of difference between gorillas and chimpanzees, which feed mainly

They use no tools

on fruit and are less fond of leaves and shoots. Schaller noted about a hundred different species of plants eaten by gorillas. Following the tracks of a group of gorillas in the Kisoro district, I myself sampled all the plants they had eaten en route. They tasted predominantly bitter, but probably met the animals' water requirements. Free-ranging gorillas have never been seen to drink, although they will wade through streams a foot or two deep. They do not cross deeper and broader watercourses unless they have been bridged by a fallen tree. Like man and the other species of anthropoid apes, and unlike almost all the lower apes and almost every other animal species, gorillas are not born swimmers. They have often drowned in the moats surrounding zoo enclosures.

Except when lying down to rest, gorillas spend the entire day eating on the move – not that they ever gobble their food or squabble over it. Because the leaves and plants which form their diet are not very nourishing they have to ingest them in large quantities. The solids in a single day's dung weigh two or three pounds, so the gross daily volume of excrement must be very considerable.

In contrast to chimpanzees, free-ranging gorillas have never been seen using sticks or other things as tools. They also show little interest in unfamiliar objects. On one occasion Schaller's rucksack was lying in full view of a black-backed male only 5 yards away. Having glanced at it once, the animal took no further notice. A passing group of gorillas showed equally little interest in a piece of white paper which contrasted strongly with the green of the forest and had been put there as a route-marker. Similar behaviour was observed on other occasions. The gorilla's personality is withdrawn – introverted, as the psychologist would put it – not only in captivity but in the wild as well. We human beings would probably live together in a far friendlier and more peaceful manner if we were more closely related to the gorilla than to the chimpanzee.

Where the senses are concerned, gorillas can see, hear and smell about as well as human beings. They rise from their home-made nests between 6 and 8 a.m., feed for about two hours and then rest between 10 a.m. and 2 p.m., interrupting their rest-period for an occasional brief snack. Shortly before nightfall, or at about 6 p.m., they start work on their padded nests. Local custom plays as large a part in nest-building as it does in diet. Plants which form part of the staple diet of gorillas living in a mountain range are never touched by gorillas living elsewhere, even though they may grow there too. Gorillas captured in adolescence or adulthood take a long time to accustom themselves to eating bread, bananas, or unfamiliar plants. Baby gorillas are less fastidious in this respect.

As regards nests, 90 per cent of those in the Kabara district are built at ground-level as compared with only 20 per cent in the Utu area. The gorillas select a spot on the ground where they can bend all the saplings and bushes inwards to form a springy mattress. They adopt a similar procedure in tree-tops or forked branches. It goes without saying that a gorilla's nest can easily support a human being. Chimpanzees' nests look similar but are always built at a far greater height and never on the ground. The gorilla builds a new nest every night, very seldom utilizing a previous site. Nests are situated a few yards apart, and there are no firm rules governing the relative position of females, children or group-leader within the camp. In the Utu and Kisoro districts, nests were sometimes located as little as sixty or seventy yards from the nearest human habitation. Incidental note: gorillas do not snore.

Silver-backed males very seldom build their nests in trees and almost always sleep on the ground. Gorillas are terrestrial creatures, not arboreal. Adult animals do climb trees, but very cautiously and only for the purpose of gathering food or, at most, in order to gain a better view of something. Females are observed in trees twice as often as males, young gorillas four times as often.

Female gorillas give birth once very $3\frac{1}{2}$-$4\frac{1}{2}$ years, but just under half their young die as babies or adolescents. Within groups, females outnumber males by two to one. If we add to the total population the male gorillas which live singly or in pairs, the ratio becomes approximately 1.5 males to every female. The mortality rate is probably higher among males. Average population density

Sun-bathing

in the relatively small areas in the Central African highlands where gorillas do occur is one gorilla to 250-1,000 acres. The gorilla is far from being a creature which only feels at home in the gloom of the jungle. Although gorillas live in forests and – at least where mountain gorillas are concerned – in regions of mist and rain, they do not avoid the sun. On the contrary, they display obvious pleasure when it comes out. Many of the animals observed by Schaller lay on their backs in the full glare of the sun for two hours or more, with sweat beading their upper lips and streaming down their chests. So far from shunning the heat, they actually left their resting-places and moved to patches of ground which were warmed by the sun's rays.

Gorillas show no enthusiasm for rain. They very often get up when rain starts to fall and shelter under trees, clustering round the trunks to keep dry. Just as often, however, they simply sit it out. Light showers do not worry them, though they usually stop gathering food. If rain starts to fall when they are already resting they turn over on their bellies or sit up. The start of a heavy downpour is the signal for young to rejoin their mothers and for any gorillas which happen to be aloft in the trees to climb down. Animals which remain seated in the open prop their chins on their chests and either fold their arms or clasp the right shoulder with the left hand. They create a most pitiable impression as they sit there, motionless and silent, with rain streaming from their shoulders and eyebrows. Hardly anything disturbs them when they are in this state. Schaller once walked straight through a group seated in this fashion, and only one of the animals raised its head. On another occasion he moved directly into a group's field of vision and perched on an overhanging branch at distances of 10-30 feet. They looked at him but did not move away. It poured for over two hours before they finally decided to gather food despite the rain. Gorillas never abandon their nests on account of rain, nor do they behave any differently under heavy falls of hail.

The gorillas observed by Schaller took no notice of thunder, which is frequent in the mountains, though they were just as startled as their human neighbour by two particularly heavy claps. One of the males threw up its head and looked at him.

Endowed with a self-assurance born of size and strength, gorillas take little notice of other creatures. Mosquitoes and other stinging and biting insects do not occur in their habitat, at least in Kabara. They sometimes jumped and swore when ravens or other birds swooped on them unexpectedly – a reaction shared by the observer. Once, a tree dassie the size of a rabbit popped out of a fallen tree-trunk on which a gorilla was sitting. Although a whole group of gorillas saw the animal, all ignored it. Elephants and gorillas, too, seem to take little notice of each other, although gorillas do avoid hill-sides which have been trampled and browsed by elephants. Groups of chimpanzees and gorillas sometimes frequent the same areas without fighting and may also sleep a short distance apart. Gorillas seem to be just as untroubled by the big red forest buffalo.

It is often stated in books that leopards carry off baby gorillas, especially at night. Observations made at Kabara did not support this. Only one young gorilla vanished from a group during the entire period, and the reason for its disappearance was never established. The dung of leopards in this district contained virtually nothing but the hair and remains of duikers, or small forest antelopes, and tree dassies. In the Kisoro district of Ruanda, a far smaller area and one where the forest is being progressively whittled down and cut back by farms and villages, the duiker population has been greatly reduced by hunting and poaching. The bodies of two gorillas were found there, one a female and the other a male which had been killed in the course of a fight with a leopard. The leopard had surprised the sleeping gorilla in its nest, and the two animals had rolled down a slope locked in mortal combat. On another occasion, a black leopard was observed stalking a lone male gorilla which was sitting there unaware of its danger. The gorilla eventually moved off before the big cat had edged close enough.

The gorilla's behaviour towards man is equally pacific. While in the forest, Schaller unexpectedly encountered lone gorillas on no less than three occasions. Surprise was mutual. Although separated from Schaller by distances of only 5-8 feet, the gorillas – two females and a male – made no move to attack him. One of the females screamed and retreated about 10 feet but the other two animals simply glanced up and then took no further notice of Schaller as he slowly withdrew. Most groups happily remained at a distance of only 20-25 yards from the observer for hours on end, although he stood there quite openly.

Groups of gorillas which encounter human beings back off peacefully and disappear. This is their normal mode of behaviour. They inflict injury only when interfered with, that is to say, when people try to kill them, surround a group and capture individual members of it, or – in rare cases – drive them off fields and plantations. When native hunters have killed a silver-backed male they surround the females and club them to death. On many occasions the animals do not even try to run away. It is pitiful to see them shielding their heads with their arms in an attempt to ward off the blows – an appeasement posture which they maintain to the end.

Nine cases of wounding by gorillas were treated at the Kitsombiro mission hospital, near Lubero, between 1950 and 1959. Only six of these necessitated a lengthy stay in hospital. Three of the African patients had been bitten while attempting to encircle and kill solitary silver-backed males. The wounds were in the thigh, calf, and hand. On another occasion, hunters cornered a lone black-backed male. One of the men slipped as he tried to run away. The gorilla caught him by the knee and heel and bit him in the calf, ripping off a piece of muscle 7 inches long. At Kayonza a few years ago a Bantu was bitten in the hind quarters. It was a long time before he could sit down, and the Batwa still laugh as they tell the story. Most of the gorillas which attack in self-defence are males, but females sometimes do so too. Tribesmen in the Cameroons regard it as a disgrace to be wounded by a gorilla because everyone knows that the animal would not have attacked if the man had not panicked and run away.

For all its superior strength, the gorilla is a peaceful and tolerant creature, not a malevolent jungle monster. The picture painted of it by big-game hunters is a slanderous fabrication, possibly engendered by their own sense of guilt. The man who shoots one of these big anthropoid apes must be emotionally affected in much the same way as if he had shot another man. After all, gorillas bear a disquieting resemblance to man in respect of stature, facial expression, behaviour, and quality of gaze. The best way of salving one's own conscience and justifying oneself to other people has always been to represent an opponent as criminal, bestial and murderous, or – at best – as an unprovoked aggressor who must be shot in self-defence. We have only to think of our own wars. Do we really have the right, and can we morally afford, to eliminate a creature like the gorilla from the face of the earth?

14. *The beast that lived like a man*

All that crows was once an egg.

AFRICAN PROVERB

In 1953 I conducted a census of all the gorillas resident outside Africa. (They numbered 56, but by 1967 their numbers had risen to 302 lowland and 12 mountain gorillas.) At that time, sixteen years ago, I heard of a woman who kept a full-grown gorilla at her home in Havana. Being aware of what it means to handle adult gorillas, I was fascinated by the information. Although I was not immediately able to correspond with the legendary lady from Cuba, I did discover in a roundabout way that the animal was female and had first come into her possession as a tiny baby in 1932.

Since then, I have often exchanged letters with Mrs Maria Hoyt, the lady in question. When E. Kenneth Hoyt married his young Austrian bride in 1926, he gave up his business interests in South America and set off with her on a protracted tour of the world which took the couple to Ceylon, Algeria, Morocco, Indochina, Japan, Europe, Hongkong, and the whole of East Africa. The Hoyts never spent longer than six weeks in the same place in any one year – never, that is, until they visited West Africa.

West Africa was where Maria Hoyt met her Waterloo, so to speak. In 1932, she and her husband left the Ivory Coast by truck for a tour of what was then French West Africa, the Sudan, Nigeria and Tchad. They were eight months en route and their trip took them over 8,000 miles of appalling roads. Having been an enthusiastic breeder of fighting-cocks in his youth, Kenneth Hoyt made the entire journey accompanied by ten laying-hens acquired in the

Ivory Coast. When released from their cages at evening halts, the hens usually made a dash for their laying-box almost before it was unloaded. If they were really in a hurry to lay their eggs, the top of a Thermos flask or some other receptacle had to be proffered for their convenience. Archie, the cockerel, put every local village cockerel to flight. Being of African stock, the hens had a remarkably keen awareness of potential danger. They swiftly and fearlessly pecked to death poisonous spiders and scorpions – even small snakes. They also clucked a warning if larger enemies were in the offing. The servants, who had been ordered to investigate the cause of any commotion among the hens, often found dangerous snakes hanging from branches or peering from under crates.

In French Equatorial Africa, now Congo-Brazzaville, Kenneth Hoyt went gorilla-hunting. He had promised the Museum of Natural History in New York a large gorilla for display purposes. Those taking part in the hunt included not only the Hoyts' fourteen servants and two drivers (one European and one African) but also at least two hundred beaters from neighbouring villages. The natives had already cornered a family of gorillas on an island in the middle of one of the large rivers in the area. The animals were then driven towards the guns. 'Close on the heels of a female with her baby, a huge male gorilla burst through the undergrowth,' wrote Mrs Hoyt. 'There was something so splendid and noble about him that I feared for his safety. My heart beat wildly and painfully as he thrust the heavy branches aside with a single sweep of his mighty arm and disappeared. Then I heard a shot, and almost immediately afterwards Ali, my husband's gun-bearer, ran up to me shouting and waving his arms. "Memsahib," he shouted, "the big gorilla is dead – finished!"'

There was worse to come, unfortunately. Unbeknown to the Hoyts, the natives had erected strong nets in the forest and driven the entire family inside. In addition to the one specimen required, eight gorillas had been speared to death. The only survivor was a baby gorilla weighing 9 pounds which furiously bit and scratched the bare chest of the headman who was carrying it. Mrs Hoyt

A pediatrician for orphan Toto

stretched out her arms to the poor little thing, whereupon it fled into her embrace and immediately calmed down. After such carnage, Maria Hoyt felt obliged to save at least one member of the massacred group, which lay there in a grisly heap. The little creature could not even walk yet, but a full-breasted young native woman volunteered to wet-nurse it during the Hoyts' last few weeks in Africa. Then one of their African servants, a youth named Abdullah, decided to abandon his three wives and accompany them back to Europe as the animal's keeper.

The Hoyts' first stop was a hotel in the Rue de Rivoli in Paris, where little Toto developed a severe case of pneumonia. A pediatrician hurriedly injected her leg with oxygen so that it swelled up like a balloon, a portable X-ray apparatus was brought to the hotel to photograph her respiratory tract, an oxygen-tent was set up, and numerous doctors conferred daily in an adjoining room. The animal's life was saved, but she became so spoilt in the process that Mr and Mrs Hoyt had to take it in turns to sleep with her.

The baby gorilla's foster-mother took her to recuperate at Arcachon on the coast not far from Bordeaux, still accompanied by a retinue five strong. Mrs Hoyt found it difficult to go for walks on the beach with Toto because they were constantly surrounded by inquisitive onlookers, mostly women. It was hard to say which attracted them more, the baby gorilla or Abdullah, who was an extremely handsome young man. Abdullah enjoyed such universal success with the hotel chambermaids that Mrs Hoyt had to warn him against vengeful boy-friends and husbands. Eventually she moved into a private apartment with a balcony where little Toto could benefit from the air and sunshine without being pestered by sightseers. Toto waxed more and more critical of the clothes her foster-mother wore. Many hats she liked and approved of, tapping them delicately with her fingers. Others were less to her taste. She not only whipped them off Mrs Hoyt's head with lightning speed but usually reduced the latest Paris creation to a torn and crumpled wreck in a matter of seconds.

Sailing to Cuba

Counselled by experience, Mrs Hoyt gave her swarthy infant a sedative before sailing with her to Cuba, so by-standers and ship's crew assumed that it was a human baby which Abdullah carried aboard in the carefully shrouded baby-basket. Having booked an entire suite and three additional cabins, Mrs Hoyt always managed to switch Toto from one cabin to another before the stewards could catch sight of her. The ship sailed to Havana, where Kenneth Hoyt had meanwhile found and purchased a villa. He and his wife felt that Havana's tropical climate would be the best possible thing for their foster-child's delicate constitution. The villa can hardly have been a very modest establishment because it now, since expropriation by Fidel Castro, houses the Soviet Embassy.

This was where Toto grew up. Much of what she did as a youngster came as no surprise to me because my wife and I have reared several baby gorillas in our own home of recent years, though not to the size and age attained by Toto. Mrs Hoyt must be unique in having kept a gorilla for so long.

Toto enjoyed playing with hard rubber balls. She also sat for hours in front of tiled walls and chalked peculiar shapes on them. Sometimes the shapes suggested numbers, in which case the gardeners and domestic servants would promptly buy lottery tickets to match. They even won on one occasion. Mrs Hoyt's mother taught Toto to draw in the air with her fingers: a circle for a face, three dots for eyes and nose, and a line for the mouth. Toto often greeted her by describing this pattern in the air with an outstretched forefinger, as though it had become the recognition signal of a secret society. She later transferred these drawings to the tiles with chalk. Although she never succeeded in reproducing them very skilfully, they were clearly recognizable as human faces.

The baby gorilla adored anything cold. At two years of age she developed a special passion for an old-fashioned ice-box which stood outside the kitchen door on the servants' veranda. It was not the food she was after, but the blocks of ice. She waited until no one was about, closed the kitchen door, propped a chair against it and sat on it so that nobody could get out, and then reached into the ice-box. She often carried off blocks of ice weighing ten pounds or more, concealed them in the garden or smashed them on the tiled floor. Although free-ranging gorillas never eat food of animal origin, Toto liked meat even when it was bloody. She retained this predilection long after she had grown up.

Toto would only tolerate the presence of Mr and Mrs Hoyt, Mrs Hoyt's mother, Abdullah, and one of the housemaids. Anyone else who tried to approach her she bit and scratched. Once, when she had behaved particularly badly, Mrs Hoyt gave her a gentle smack. Toto promptly uttered a terrible cry and flounced off into a corner, where she turned her face to the wall and started to stamp and scream with such violence that the entire household came running to see what had happened. Mrs Hoyt's heart sank at the prospect of Abdullah's forthcoming return to Africa.

There was no alternative. The colonial government insisted that every servant taken abroad should be shipped home at the employer's expense after a maximum of two years. The Abdullah who embarked at Havana bore little resemblance to the gun-bearer in khaki shirt and shorts who had left Tanganyika with the Hoyts two years earlier. He now travelled with three large trunks full of shirts, ties, suits, underwear, shoes and hats, not to mention a host of presents for his wives and friends. All in all, he looked less like a young African than a successful Harlem band-leader. News soon arrived from Tanganyika that Abdullah had disposed of his entire European wardrobe at vast profit immediately on arrival. What was more, his three wives had wisely invested the money which he had been sending home each month. He used it to start a business, and soon owned butcher's shops in three different villages.

Mrs Hoyt ran through five new keepers in quick succession during the weeks that followed Abdullah's departure. One was a Cuban woman who tried to amuse the young gorilla by rattling castanets in her face. Toto eventually became so furious that she wrenched them out of the woman's hands and smashed them over her head. In the end, Mrs Hoyt was fortunate to find an efficient young Spaniard named Thomas, who looked after Toto for more than thirty years and grew old in her company.

Thomas had previously been employed by a woman named Abren, who had a special affection for anthropoid apes. She was the first person to breed chimpanzees in captivity and cherished the belief that these manlike creatures were endowed with an immortal soul. She actually built a chapel on her property and attended Mass with her charges. When she died in 1930 her son presented her large collection of monkeys to various zoos, so Thomas was out of a job. He quickly lost his heart to the baby gorilla. Year in, year out, he never left her side except for an occasional hour in the afternoon. During the first six years they never spent a night apart, and for a while they shared a bed in the new house that had been built for the animal.

Sometimes in the middle of the night, Toto would feel homesick for Mrs Hoyt or her husband. Slipping secretly out of bed without waking Thomas, she quietly opened the door, closed it from the outside, and made for the couple's bedroom.

Later, a special iron bedstead was made for her. This had a barred top which enabled her to be shut in for the night. The bed also had springs and a mattress, but a hole had been cut in one corner of the mattress and beneath it stood a chamber-pot. Unlike many human children, Toto learnt to keep her bed clean straight away. She used the same pot during the day-time. Because she quickly found out how to turn on taps and drink from them – and the Cuban water was not without its impurities – all the taps had to be modified so that they could only be operated by means of a square key. The Hoyts regularly sent stool and urine samples to a medical institute for testing over a period of some years. The samples were simply marked Baby Toto, and none of the laboratory staff ever suspected that they were not of human origin.

Toto and Thomas always ate together, their meals being brought to them on two separate trays. Toto got very annoyed if Thomas's meal arrived first, and gobbled it up. If the trays arrived together, she would carefully inspect her keeper's to see if there

237

The cage bed

was anything on it which did not appear on hers. If so, she snatched it away and sampled it to see if she liked the taste. For this reason, it soon became an established rule that Thomas and Toto should be brought identical meals. Toto ate with a spoon throughout her life.

By the time Toto was three she had the strength of two grown men and enough mischief in her for a dozen growing boys. There was not a door in the house that she could not open. It was inadvisable to lock one because she would search and search until she found the key or simply wrench the handle off. When she became older and heavier she quickly discovered that any door would give way if she threw her weight against it.

A broad flight of steps led from the villa into the garden. It was bordered by marble balustrades, each of which supported marble columns half-way down and at the foot. One day Toto suddenly raced up the steps, clambered on to one of these broad marble balustrades, and slid down at top speed. It looked as if she would collide with the central column, but she took evasive action. Flinging one arm round it, she sailed through the air in a semi-circle, landed on the lower section of balustrade, repeated the manœuvre on reaching the lower column, dropped on to the marble paving, and completed the performance by turning half a dozen somersaults. Henceforward, this became her favourite sport on the occasions when she was allowed into the villa. She also somersaulted the entire length of the steps from top to bottom. Even when shut up in her bed, which two men could not lift, she managed to propel it across the room to the bell-push and ring for servants.

The house that was eventually built beside the garage for her personal use consisted of a play-room measuring $16' \times 26'$ and a $12' \times 16'$ bedroom. Adjoining it was a barred enclosure measuring $40' \times 90'$, but she was only locked up there when the Hoyts had visitors of a nervous disposition. The door was always open at other times, enabling Toto to roam as she pleased. Thomas began by placing his own bed next to hers in the sleeping area. Later, over a period of several months, he moved it inch by inch towards the play-room until he managed, little by little, to close the connecting door. In the end he was able to sleep on his own without provoking an outburst of crying and screaming.

One day while Thomas was in her bedroom, Toto sneaked out and bolted the door behind her. Having safely locked Thomas in, she danced up and down with glee in the outer room. It was an hour before she tired of the game and obeyed Thomas's summons. He scolded her through the door and ordered her to unbolt it, which she duly did.

In view of her predilection for tearing things up, Thomas used to give Toto large palm-fronds and branches to play with. Nothing was safe from her attention unless she was kept occupied. Next to the garage was the laundry. One of the gorilla's pet tricks was to grab a shirt or handkerchief when no one was looking and rip it to pieces. There was nothing for it but to rail off the drying area and put bars over the doors and windows of the wash-house. Even then, she occasionally managed to unbolt one of the barred doors. Servants ran screaming in all directions, whereupon Toto took over. She pulled everything to pieces and continued to rampage until Thomas arrived with a stick and took her back to her own quarters.

Toto was extremely docile and well-behaved for most of every month, but there were two or three days when she became excitable and – as her strength increased – dangerous. According to Mrs Hoyt, her awkward days always coincided with the new moon. Personally, I suspect some connection with the menstrual cycle. Mrs Hoyt assumed that gorillas become sexually mature at the same age as human beings. We now know that they do so at the age of six – yet Mrs Hoyt let her jungle guest roam free until she was nine years old!

Sometimes during such periods, Toto played hide-and-seek with the members of her human family and made them comb the garden for her for hours on end. She would conceal herself somewhere in the dense undergrowth and crouch there, quiet as a

The gorilla understood Spanish

mouse, until one of the search party almost stumbled over her. When someone found her and called for Thomas, her immediate response was to take off in a different direction and find a new hiding-place. 'Because, although she couldn't speak, she understood Spanish as well as any Spanish child of her age. She couldn't read, fortunately, so we made a set of placards. All we had to do, once we had found Toto, was to hold one of them up so that Thomas would know where to go.' It was not long before Thomas himself grew reluctant to follow her on her excitable days. She used to throw sand and gravel at him, and once, in her play-room, grabbed a swing and sent it careering at his head with all her strength. The only thing that chastened her was a stick which administered mild electric shocks.

One day Thomas and Toto found four kittens under a bananatree at one of their favourite spots in the garden. Their mother was a cat belonging to the villa. Toto was enchanted. She examined the little creatures carefully, smiled, made 'mm, mm' noises with her lips, and touched the kittens gently with her fingers. Finally she picked up a little black-and-white kitten and cradled it in her arms. The kitten purred delightedly and snuggled against the gorilla's thick soft fur.

Toto was so preoccupied with her new baby that she had no time for anything else. She could not even be persuaded to dance to the gramophone, which she normally did with great pleasure. On the way back to her quarters she carried the kitten carefully on one arm and walked on her hind legs and the knuckles of her right hand.

Henceforth, Toto took Blanquita with her wherever she went, sometimes under one arm, sometimes on her back, sometimes draped round her neck. Even when Blanquita had grown into a cat, she spent almost all her time with Toto. Then, one day, Blanquita herself gave birth to six kittens. Toto selected one of them – a little male named Principe – and from then on showed little interest in the mother. She barely glanced at Blanquita when the cat came and rubbed against her, and often pushed her impatiently away.

Toto had another good friend in the shape of Wally, a bull terrier bitch. Wally was the only one of the Hoyts' dogs which never developed a fear of Toto. When Toto was still a baby she used to romp on the lawn with all the dogs, biting, scratching, and rolling around in a way which delighted the dogs as much as it did her. Later, when she became bigger and stronger, she could stand erect with a dog in her arms and throw it a considerable distance. Her canine friends gradually became nervous of her, but not Wally. Toto and Wally were inseparable, and loved one another dearly.

One day the dogs started to fight in earnest. The gorilla was only moderately interested at first, and simply moved closer to watch. Then she discovered that Wally was involved, getting the worst of it, and being bitten by two other dogs. Toto sprang to life. Uttering a deep throaty cry, she went to her friend's assistance.

Thomas ordered her to stay where she was, but she refused to obey. When he held her back she promptly bit him in the arm and scratched him in her furious determination to fight shoulder to shoulder with Wally. Thomas eventually managed to drive her back to her quarters with the aid of a stick and lock her in, but she did not forget the incident. Two days later, when she was walking in the garden with Thomas, she met one of the dogs that had been biting Wally. Quick as lightning, she flew at the animal with her teeth bared. Murder would have been done if Thomas had not made a superhuman effort to prevent it. From then on, the two canine trouble-makers had to be locked up whenever Toto herself was not confined to her quarters.

When Toto was young and her first milk-tooth worked loose, she came to Thomas with her mouth wide open and one finger on the tip of the little tooth. She waggled it back and forth to show how loose it was and gave him a look which eloquently demanded the reason. Thomas took a piece of string, tied it round the tooth, and pulled it out. Then he gave her a little brandy and water because he thought it would be good for the raw socket. Toto drank

Toto drops the gardener

up and asked for more, not that she got any. Whenever a milk-tooth worked loose in future, she gaily trotted off to her keeper, eagerly showed him the tooth, and held her mouth open while he attached the string. As soon as the tooth was out she demanded brandy.

Toto loved midges. Unlike the human inhabitants of Cuba, who were plagued by them in May and June, she enjoyed almost complete protection because her fur was too thick. As soon as one landed in the region of her eye she grabbed it with unfailing good aim and ate it. She also waited with equal interest for midges to land on her human friends and treated them to the same fate.

As her strength increased, so she delighted in showing it off. One day she seized the little Japanese gardener, Kayama, and carried him kicking and screaming up the outside of her exercise cage. Six or seven feet from the ground she suddenly lost interest in the game and dropped him.

On another occasion, the master of the house was entertaining a friend who happened to be an amateur boxer. Toto was confined to her quarters for safety's sake, but the visitor insisted on shaking hands with her through the bars. Kenneth Hoyt warned him against doing so, although she was only five at the time. Confident of his strength, the boxer extended his hand and was promptly yanked against the bars with such force that he cried out in pain. The Hoyts at once drove him to hospital but his shoulder was found to be intact.

The only person whom Toto treated with unfailing gentleness was Mrs Hoyt's elderly mother, who often brought the gorilla her morning snack. Toto always took her hand, kissed her and laid her arm against her ear so that she could hear the ticking of her wristwatch. Thomas, her keeper, could never afford to wear a watch because Toto invariably removed it from him and smashed it. Perhaps she felt that only the dear old lady with the white hair was entitled to something which looked and sounded so nice.

Towards the end of the gorilla's life, a whistle-signal was introduced. If the whistle rang out, it meant that Toto was loose. Any-

Mrs Hoyt out of action

one who was afraid of her – and she weighed 4 cwt. by this time – took cover with all possible speed. Doors were barricaded and windows closed. For a while, Toto savoured the heady sensation that she was mistress of all she surveyed. Then she would scamper up to the villa, rattle the doors, climb up the grilles covering the windows and peer in. Her keeper generally defended himself with sticks and shields cut from thorn-bushes. He never threatened to resign his job even though she often roughed him up and repeatedly tore his clothes, but the position became more and more impossible.

In December 1937, while Toto was playing with her foster-mother, she gleefully leapt on to her swing and sailed into Mrs Hoyt, knocking her to the ground. The poor woman fell backwards on to the flagstones and broke both arms. The gorilla was terribly upset when she saw that Mrs Hoyt was badly hurt, and rained kisses on her. Realizing that her husband would get rid of Toto if she told him what had happened, Mrs Hoyt pretended that she had slipped and fallen. She could tell, however, that he did not really believe her.

It was three months before Mrs Hoyt regained the full use of her hands. Toto looked quite cast down whenever she encountered her. Taking Mrs Hoyt's hands gently in her own, she turned them over with great care and blew on her palms and kissed them, just as Mrs Hoyt used to do to her when she had grazed or scratched her hands as a baby.

Shortly afterwards Kenneth Hoyt became gravely ill and died in a New York hospital. Mrs Hoyt, who felt dreadfully bereft, found it more and more difficult to keep the massive gorilla on private premises. When Toto was eight, her strength was such that nobody could compel her to do anything against her will. In addition, there were two or three days in every month when she appeared to fall in love with one of the men in the house. Sometimes it would be one of the gardeners, a tall and handsome young man, sometimes José the second chauffeur, and sometimes the butler. On such days she used to trail the object of her affections all round

Toto falls in love

the grounds or sit down and stare fixedly at him while he worked. She also tried to touch him, though Thomas and Mrs Hoyt did their best to prevent this because they never knew what it might lead to. Toto grew annoyed with them if they tried to lure her away from her current light-of-love or distract her attention. She would pelt them with flowers, of which she was normally very fond, and angrily thrust their hands away if they tried to caress her.

Sometimes Toto would forget her affairs of the heart and roll happily on the lawn like a big Teddy-bear. Then, without warning, she would rise to her full height and charge like a maddened bull. It all happened so quickly that there was no time to run away. The only course was to fling oneself to the ground and roll aside. This appeared to satisfy her, because she usually called off the attack. However, there were other times when she would seize her foster-mother by her dress or one arm and drag her for yards across the lawn, along a path or through her sand-pit, not releasing Mrs Hoyt until Thomas had come running in response to her cries for help. These attacks exactly resembled those made by free-ranging mountain gorillas at Kiwu. They, too, seldom continued to bite a female or an adversary once the latter was lying on the ground in an appeasement posture.

Toto did not abandon her old games. She used to take Mrs Hoyt's hand and guide it to her ribs or the soles of her feet to show that she wanted to be tickled, then smile all over her face when her wish was granted. She also indicated when she was to be brushed, or lay down with her head cradled in her foster-mother's lap and promptly fell asleep. Her pet habit was to pick pockets for keys and conceal them, usually in the fold between the top of her thigh and her big jutting belly. She could even walk with a key hidden there and not let it fall. Having hidden something of the kind, she would open her mouth, stick out her tongue, spread her hands wide and raise her arms to show that she had nothing on her. Only earnest entreaties or a good scolding would persuade her to surrender the article.

Sometimes the gorilla herself noticed when her mood was

deteriorating. If play became too rough, she would suddenly put her arms round her foster-mother, kiss her, and push her away from the table on which she was sitting. At other times her excitement mounted so rapidly that she would tear Mrs Hoyt's clothes off, rip them to shreds and bear them away to some roof or other. For this reason, Mrs Hoyt kept a special drawerful of spare garments in the gorilla's play-room.

In Mrs Hoyt's opinion, gorillas are incapable of any form of self-control. She believes this to be the greatest difference between them and human beings – more significant than their inability to speak. Their sole rudiment of self-discipline may be this habit of terminating social intercourse with a friendly fellow-creature as soon as they feel on the verge of being overwhelmed by agitation or rage.

Toto had an aversion to sunlight and always sought out patches of shade. This was why she gradually developed a hatred of photographers. Because gorillas have dark fur, would-be photographers always wanted her out in the sunlight. Hence the many photographs of her which show the tip of the electric stick held by Thomas, who had to deter her from attacking the photographer and his little clicking box.

In 1940, when Toto was once more running amok in the grounds and terrorizing all and sundry, Mrs Hoyt only succeeded in getting her back into her cage by means of a trick. Telling the chauffeur to take her by the shoulders and shake her, she shouted: 'Toto, Toto – help me!' Almost instantaneously, heavy footsteps were heard pounding over the flagstones as Toto galloped to her rescue. No sooner had she caught sight of José than she uttered a menacing grunt and charged inside like an express train, her eyes glinting with fury. José fortunately managed to escape at the last moment and slammed the door behind him in the gorilla's face. She hammered it with her fists and threw her weight against it in an attempt to break it down. Once again, Mrs Hoyt and Thomas had succeeded in getting her safely behind bars.

In 1941, after the authorities had threatened to have the large

Off to Ringling Brothers

and dangerous animal shot, Mrs Hoyt was finally compelled to hand Toto over to Ringling Brothers' circus. She did not sell her, but retained the right to take her back at any time and spend as much time with her as she wished. The big problem was how to get Toto aboard ship for the voyage to Florida. It was proposed to carry her out of the grounds shut up in her iron bed, which was reinforced with strong steel mesh. As soon as Toto realized what was afoot she flew into a rage. Although they were three-quarters of an inch thick and had safely restrained her for the past twelve years, she gripped the iron bars enclosing her bed and burst them. It took twelve men two hours to move the huge animal from her living quarters to the caravan which had been specially built for her. Once the iron bed was finally inside and had been opened up, Toto sprang out with a single furious bound and rushed at Thomas, but only to embrace him and seek consolation. She was terribly sea-sick throughout the voyage and ate nothing. Neither she nor Thomas slept, but sat side by side holding hands and consoling one another.

Mrs Hoyt, who had gone on ahead by air, got a joyful reception from her foster-child. Waiting for Toto at the circus was Gargantua, a huge male gorilla aged eleven. There was excited speculation among journalists and circus staff over how the 'bridal pair' would react to one another. Because it would have been too risky to put the animals together immediately, they were housed in the air-conditioned caravan so that they could only see and touch one another through a grille. Toto wanted nothing to do with the big black stranger. Her initial surprise gave way to fury. Gargantua had only to put his hands through the grille in quest of friendly contact, and she would stamp and roar like a mad thing. When he took a stick of celery from his meal and tossed it into her compartment as a peace-offering, she promptly hurled it back in his face.

The cat Principe continued to accompany Toto on her travels for years to come, never deserting her caravan for long. Mrs Hoyt travelled with the circus for the first seven months until Thomas had learnt how to prepare the meals that suited Toto best. During the decades that followed, she usually spent the summer in Europe or elsewhere, but always returned to Sarasota, Florida, to keep Toto company for three or four months out of every winter.

There were naturally those who said that it was a sin to lavish so much money and affection on a gorilla – that it was unfair to have adopted a wild animal in the first place. But Maria Hoyt had never *wanted* a baby gorilla in her home. She had simply been presented, all those years ago in Africa, with a decision: either to let a helpless little creature die or to look after it. She herself could not have guessed how the story would end. A gorilla as accustomed to human beings as Toto cannot be returned to Africa, where it would speedily die of starvation or a hunter's bullet. Toto died on July 28th 1968 in her big travelling-cage at Mrs Hoyt's house in Venice, Florida, and lies buried in the animal cemetery at Sarasota. Her life-span in captivity, 36 years, is the second-longest on record.

So a curious quirk of fate had closely linked two very different creatures for a period of more than three decades: a frail Austrian woman, and a huge 4 cwt. gorilla from the Congo.

Mrs Hoyt wrote me the following description of how Toto met her end.

'At the beginning of March 1968 Toto began to suffer from constipation, although this was just the reverse of her usual condition – she had always been very prone to diarrhoea throughout her life, and we always had to adjust her diet accordingly. When she suddenly became constipated, the doctor advised me to give her orange juice containing "Siblin", a water-absorbent substance. It helped a bit, but we had to follow it up with doses of castor oil. She was still cheerful, though, eating as usual and showing no other change of any kind. Then, when I visited her one day in May (as I did every day), I suddenly noticed that she was not using her right arm and foot, and that she moved awkwardly. I immediately thought of a stroke, and still believe that this was what caused her death. She also developed the excessive appetite which has been observed in human beings. For instance, I always used to bring her a piece of vanilla cream-cake, and she always ate the

Toto likes TV

filling and threw the rest away. From now on she grew annoyed if I tried to feed her the filling with a spoon, demanded the whole slice and ate it at a gulp. She also ate grapes whole, although she had always sucked them dry and spat out the skins. In general, she wanted more to eat and drink than she had ever done. Her mobility gradually deteriorated. We strung ropes in every direction so that she would have something to lean on and cling to wherever she went. In spite of everything, she was always very affectionate and kissed me daily when I arrived and departed. Sometimes she took my hand, rested her head on it and fell asleep. She was still interested in her television set, and became annoyed if anyone switched it off. I spent the whole of every afternoon with her. I brushed her

Page 249:
'White' elephants occur at Etosha National Park in South-West Africa. It is startling to come across whole groups of these white animals in the bush. In fact they acquire this coloration by wallowing in the mud of the Okerfontein watering-place. The soil there is white, and clings to the elephants' hides. 'Red' elephants occur for similar reasons at Kenya's Tsavo Park in East Africa.

Page 250, above:
The white rhinoceros is the third-heaviest land animal in the world, being exceeded in weight only by the two species of elephants. White rhinos can attain a shoulder-height of almost 6 feet and a length, from tip of nose to base of tail, of $13\frac{1}{2}$ feet. The danger of extinction, which was almost total in 1920, has been banished by the efforts of conservationists. I photographed these specimens at the Mkusi Reserve in Zululand, South-West Africa.

Page 250, below:
White rhinos enjoy mock battles in cooler weather, though bulls sometimes fight furiously and in real earnest. On one such occasion, 43 other rhinos gathered round to watch.

Page 251, above:
At Voi in Tsavo National Park, Kenya, orphaned baby elephants and rhinos are bottle-fed. Even when they are bigger, they visit the game warden's post every evening or allow themselves to be driven there and shut up for the night. They are completely tame with human beings, me included.

Page 251, below:
Dwarf mongooses or kitafes are small East African carnivores which live in groups and often occupy termites' nests or hollow trees. They spend all their time scraping away leaves and grass in search of insects. I have reared several babies of this species as pets. They become very affectionate but can be a nuisance because of their habit of searching pockets, crawling up sleeves and trouser-legs, and generally 'cleaning up'.

How Toto died

hair, which remained wonderfully glossy to the day she died, and washed her face and hands, etc. She loved cleanliness and often turned round, very laboriously, so that I could wash her – she never liked being dirty.

Whenever she heard me coming (which she did from a long way off) she used to bang on the door where I always sat with her. On the last day, July 27th, she hadn't the strength to bang any longer. I noticed at once how her hand trembled when I handed her the usual piece of cake, but she did later eat a lovely big mango and drank her milk before going to sleep. When she didn't bang, I heard her making kissing noises with her mouth to greet me! You can't imagine, my dear Professor, how close to me she was and how very much I miss her. At 10.30 a.m. on July 28th she lay down with her head on her pillow and one hand beneath her cheek, and fell asleep for ever.'

Page 252:
Max, the first gorilla to be born in Germany. His mother was Makula, who came to us as a tiny baby and was reared in my home for several years. Unlike the majority of the lower apes, gorillas do not engage in mutual delousing. Although young chimpanzees and baboons are only too pleased to undergo fur-cleaning by their mothers and often invite their attentions, baby gorillas resist like bath-shy children. Gorillas have twenty-two different sounds in their vocabulary. They yawn, snort, cough, belch, hiccup and scratch themselves like human beings.

15. Why don't camels die of thirst?

The words of the old are like the dung of the hyena, dark when fresh, then light.
AFRICAN PROVERB

Beside Lake Rudolf in Kenya, in Black Africa proper, I came across large herds of dromedaries.

Any boy who has read Karl May will know how Kara ben Nemsi and Hadji Halef Omar saved themselves from dying of thirst in the desert. They slaughtered one of their camels, slit open its belly, and drank the store of water which it allegedly carried in a peculiar pouch next to its stomach. This iron ration of water is also reputed to solve the mystery of how camels can trek through the desert without drinking for days and weeks on end, when horses and human beings would long ago have perished.

Books of adventure contain miraculous tales of how far and how fast a camel can travel. A trained saddle-camel once covered a record 250 miles in five days with a rider on its back – 50 miles a day – though this was in 'winter', when it is not inordinately hot even in North Africa and the Sahara, and when the plants on which camels graze are reasonably green and succulent.

Caravans travel at a leisurely speed of $2\frac{1}{2}$ miles per hour. The animals have to be rested from time to time, so they cover upwards of 12 miles a day. When horses are raced against camels – which in Africa are always single-humped dromedaries – the horse usually wins over shorter distances and the riding-camel over distances that take several days to cover, though this depends on what horse and camel are expected to do in individual cases by those who bet on them.

Inside a camel's stomach

Recent books on natural history make little mention of an emergency reserve of water in the camel's belly. Anyone who takes the trouble to watch a North African camel being slaughtered and disembowelled will certainly find that it contains a porridge-like mass, but this contains less moisture than the stomach-contents of a cow or other ruminant. One can, of course, strain the liquid through a cloth, but it has a salt-content comparable with that of blood and does, in fact, consist of digestive juices, not water. A man would really have to be at death's door to sample this greenish and vile-tasting soup. The German-born Carl Raswan, who lived for many years with the Rualas of Central Arabia, was once reduced to such a pass during an armed raid in which he participated. The precious horses were closer to death than their riders. 'In order to provide our mares with something to drink, Rashid had fourteen spare camels slaughtered. The stomachs and intestines of these camels supplied enough liquid to fill eleven water-skins. When strained through herdsmen's cloaks and mixed with ten litres of milk yielded by a few she-camels, this peculiar beverage was found drinkable by our mares... With blood-spattered beards and dishevelled locks, the slaughterers bent over the carcases of the dead camels, plunged their bare arms feverishly into their entrails, tore out the stomachs, and poured the sour-tasting fluid into the water-skins...'

Whether or not a man or horse can really drink the contents of a camel's stomach, the camel itself cannot consume more liquid than it has previously absorbed. Human beings can also go without drinking entirely in cold weather provided they eat enough juicy fruit and fresh vegetables, so it is hardly surprising that the camel requires no water for months during the North African winter. On the other hand, the fact that it can survive ten times as long as a man and four times as long as a donkey in the blazing heat of the desert summer is quite enough to merit a scientist's curiosity.

All land creatures, from the little desert rat to the two-legged camel-driver, consist mainly of water and lose their body-fluid in accordance with precisely the same laws. They lose it via their

Is the hump a chemical water-store?

kidneys because they have to expel uric acid and salts, via their lungs while breathing, and via their epidermis and oral membranes because they constantly evaporate moisture so as to keep cool and maintain an even body-temperature. The kangaroo-rat cuts down the fluid in its urine to such an extent that it turns solid immediately after being excreted. When the heat increases during the daytime, the little creature crawls into its moist burrow deep in the ground, as many small desert creatures do.

The camel cannot crawl into the ground, so how does it stay alive? Recent text-books have supplied a neat explanation of this mystery: it does so with the aid of the fat in its hump. After all, the constituents of our bodies – protein, starch and fat – all contain hydrogen. When burned up and combined with the oxygen in the air, they produce water. It has been calculated that 100 grammes of body-protein will yield 41 grammes of water as a result of this 'combustion' in the body, and 100 grammes of body-fat no less than 107 grammes. A 40-kilogram hump of fat on a dromedary's back will yield upwards of 40 litres or $8\frac{3}{4}$ gallons – an obvious explanation which simultaneously solves the vexed question of why camels, and only camels, carry such humps of fat on their backs.

There is, unfortunately, one drawback to this neat theoretical solution. The animal must absorb sufficient oxygen via its lungs in order to burn the fat, but the body loses more moisture by exhalation than it gains by conversion of fat. And here the problem becomes intriguing – so intriguing that two American scientists, Dr Schmidt-Nielsen and T. R. Haupt, together with Dr Jarnum of the University of Copenhagen, were prompted to travel to the Saharan oasis of Beni Abbès, south of the Atlas Mountains, to make a serious study of the subject.

The first problem, a totally unexpected one, was to obtain some camels. Nobody wanted to sell any at first, but the difficulties attending their procurement made the scientists all the more conscientious about weighing them, testing their blood and urine, and taking their temperature.

Beni Abbès becomes unbearably hot in summer, when Europeans are seldom to be seen. The air temperature climbs to 50°C. (122F.), and the readings of 70°C. (158F.) can be recorded in places where the sun's rays are reflected by stone. At temperatures like these a man loses 1.14 litres of sweat an hour and naturally becomes very thirsty. A man who has evaporated 4.5 litres of sweat, or 5 per cent of his body-weight, finds it difficult to see and appraise his surroundings. A 10 per cent loss of body-weight robs him of hearing, causes terrible pain, and sends him mad. Human beings can go without drinking for quite a long time in cool surroundings and do not die until their body-weight has decreased by 20 per cent. In the heat of the desert a 12 per cent weight loss incurred by not drinking causes death by heat-stroke.

A camel has greater endurance. One animal which the scientists left without water for a week at the height of summer lost 112 pounds, or 22 per cent of its body-weight. It became terribly emaciated, with a shrunken belly, shrivelled muscles, and legs which looked unnaturally long. Although certainly unable to work or travel far, it did not give the impression of being seriously ill. It is to the three scientists' credit that they did not experiment to see how much weight the camel had to lose before it died of thirst. Once it had reached this stage, they watered it. It drank one bucket after another, perceptibly regaining its normal shape. Thus a camel can probably lose up to 25 per cent of its body-weight, even in blazing heat, without fatal consequences. It would certainly take far more to kill it.

One reason why camels can endure thirst so much better than human beings has to do with the fluid content of human blood. On the face of it, camel and human being both have the same amount of serum in their blood – roughly one-twelfth of the fluid content of the entire body. When a camel has lost a quarter of its body-weight by evaporation, however, only one-tenth of its serum has disappeared, so the blood is almost exactly as thin as it was. Human beings who reach this stage have already lost one-third of the serum in their blood, which becomes very thick, flows slowly, fails to penetrate the finest blood-vessels, and circulates sluggishly.

Red corpuscles conserve water

There is a consequent inability to convey the increasing heat from the interior of the body to the skin, where it is dispersed. Our body-temperature rises and we die of heat-stroke.

Another of the camel's secrets has since been discovered by the Israeli zoologist Professor K. Perk of Revohot University. Not only do camels quickly store water throughout their bodies when drinking, but absorption of fluid can dilate their red corpuscles to 240 per cent of normal size without rupturing them. This is yet another reason why camels can cope with extremes of thirst followed by a huge intake of water.

Dr Schmidt-Nielsen was also anxious to see whether his test-camel, like the kangaroo-rat, excreted an increasing volume of uric acid when deprived of water. In order to produce plenty of uric acid, he tried to feed it on a protein-rich diet of ground-nuts. He had not, however, reckoned with the animal's stubborn resistance to change. It was reluctant to give up its bad old diet and at first refused to eat at all, so the scientists had to accustom it gradually to increasing quantities of ground-nuts. Before he had properly succeeded, the oasis ran out of ground-nuts altogether and the experiment had to be abandoned.

The camel therefore has a far greater ability than man to tolerate a lack of fluid in the body, but that is not all: it also gives off far less moisture. When we are in air of a far higher temperature than our bodies (which preserve an even 37°C. (98.4F.)) we begin to sweat. Only by evaporating body-fluid do we succeed in preventing the interior of our bodies from growing hotter and hotter. But sweating, as we have seen, is correspondingly expensive in terms of water.

Things are different with the camel. During the day-time, when the sun beats down and the air becomes torrid, its body-temperature continues to rise until it reaches 40°C. (104F.) The animal does not start to sweat until this point has been reached, which naturally saves a great deal of water. At night, when the desert becomes extremely chilly, its body temperature drops rapidly to as little as 34°C. (93F.) Because of this daily 6° fluctuation in temperature, it is naturally some time before the camel's big frame warms up sufficiently for sweating to commence. The animal's body-temperature only fluctuates to this extent in summer. The variation is far smaller in winter and on the Mediterranean coast.

But that does not exhaust the list of discoveries made at Beni Abbès. Donkeys are desert creatures too, and can likewise, in contrast to human beings, lose up to a quarter of their body-weight when deprived of water. They do, however, lose body-fluid three times as fast as a camel. A dromedary has been known to go without water for seventeen days, even in the molten heat of the desert, whereas donkeys have to be watered every four days. Their body-temperature can fluctuate to a greater extent than man's, but not as much as a camel's. In other words, donkeys start to sweat far sooner. This is due, among other things, to their thin coat of hair. Although camels lose their hair in summer they retain 2-4 inches of felt on their backs and are thus admirably protected from the sun's rays. (Unlike Europeans, who don light clothes in summer, the bedouin of the desert wears woollen burnouses, often many layers of them.) Fat is a very poor conductor of heat, so it is particularly advantageous to camels and dromedaries to store their fat on their backs because it provides them with further protection from the sun. If the fat were distributed throughout the body between intestines and muscles, it would obstruct the outward flow of heat to the surface of the body.

The donkey is superior to the camel in one respect. A parched dromedary can regain its original body-weight by drinking 30 gallons of water in ten minutes – in fact it is almost frightening to watch an animal empty ten buckets in such a short time – but a donkey can put on a quarter of its original body-weight in only two minutes. Human beings who have been out in the heat of the desert for a day take several hours to restore their weight by drinking, and often have to eat as well.

The ability to drink quickly is an asset to the wild animal. Water is scarce in summer, and water-holes, which are generally few and far between, attract lurking predators. A creature which

Cars versus camels

can 'tank up' in two minutes runs a correspondingly smaller risk of attack.

The camel's mysterious aptitudes, of which nothing was known until recently, facilitated trade throughout North Africa and large parts of Asia for hundreds of years. They enabled kingdoms to flourish and extended man's sovereignty over remote and comfortless regions. In the middle of the last century, camels were imported into North America. Their ability to endure hunger and thirst again proved itself to such an extent that the animals had already begun to affect the destiny of the New World when they were swiftly overtaken by the advent of the railroad.

The same applies to the Sahara. A car does not sweat and can really secrete water-tanks in its intestines. It does not drink as quickly as a camel, but it can travel faster and farther. This makes it far superior to the good old ship of the desert – always provided that every cog in its metal belly functions as it should.

16. Too few elephants or too many?

The elephant, too, dies but once.

AFRICAN PROVERB

What is the real truth about elephants in Africa? Newspapers sometimes claim that thousands of them ought to be shot because they have over-populated and devastated entire regions. Against this, old African hands complain that one can now drive through Africa for weeks without seeing a single elephant or even a single trace of elephant-dung – that capital bull elephants with the really huge tusks coveted by big-game hunters are a thing of the past.

The golden age of the Proboscidea is long since gone. Elephants, which began to evolve fifty million years ago, settled in all the world's continents with the exception of Australia. The remains of more than 350 different species have already been unearthed. Our forebears in Europe and Asia coexisted with smaller, hairier elephants – or mammoths – of the sort which occasionally come to light in Siberia, frozen solid complete with flesh and blood and fodder between their jaws.

It is not long since the elephant's evolutionary relationships were identified. Elephants have very little connection with rhinos and hippos, with which they used, quite unscientifically, to be classed as pachyderms, or thick-skinned animals. Modern zoologists do not even classify them as ungulates (hoofed animals) proper. Instead, on the basis of anatomical affinity, they group them with the siren or sea-cow and the rabbit-sized dassie in the order referred to as subungulates or 'pre-hoofed animals'. The scientific linking of such outwardly different creatures always prompts zoo visitors to wag their heads in bewilderment.

The first zoo-born elephant

Today, only two easily distinguishable species of elephants still walk the earth: the Asiatic or *Indian elephant* and the *African elephant*. The Indian elephant is rotund and massive, with prominent frontal ridges, smaller ears, and a head which forms the highest point of the body. The African elephant, by contrast, is higher in the back. It has a receding forehead, much larger ears, is thinner, and differs from the Indian elephant in having two finger-like organs at the tip of its trunk.

Asiatic elephants are far more commonly found in zoos. The 1965 figures for all American zoos showed that only 30 African elephants were resident there, all of them imported, as compared with 91 Indian elephants of which 4 had been born in the United States. The first African elephant ever to be born in a zoo came into the world at Munich in 1943, and only a handful more have been born since then.

Men have been training Asiatic elephants to do useful work for thousands of years. The Carthaginians may well have done so with the relatively small and now long extinct North African elephant. It is probable that the elephants with which Hannibal crossed the Alps in 220 B.C. and struck such terror into the Romans were of African origin. Sixty years before, King Pyrrhus of Epirus had fought several successful battles against the Romans in Italy, also using elephants. For some two thousand years thereafter, however, the African elephant was regarded as untameable in comparison with the Indian. The Belgians showed in recent decades that this was not so. They trained African elephants to pull carts, carry riders, and do all manner of jobs. I myself spent some time at the army-run elephant-training centre at Gangala na Bodio in the North-East Congo, and have described the technique in my book *No room for wild animals*.

Generally speaking, however, the grey giants are on the retreat in Black Africa too. Huge herds of elephants were still grazing in the vicinity of Capetown when the first Dutch settlers reached South Africa, and the ivory market still derived substantial quantities of tusks from that area in the early decades of the last century.

Hunters soon changed that. Even the celebrated Kruger National Park was without elephants until a few migrated there from the adjoining Portuguese territory of Mozambique. The big animals also maintained a foothold north of the Zambezi.

Elephants, all of which probably originated as forest-dwellers, resemble the rat, human being and sparrow in their superior ability to adapt themselves to a variety of habitats. African elephants can be seen standing up to their necks in water chewing rushes and water-plants; they also climb high into the mountains, sometimes attaining altitudes of more than 16,500 feet; they graze like cows in some areas and browse on trees in others.

This is why African elephants are far from universally identical. During the last century, zoologists followed the prevailing fashion and distinguished dozens of elephant species by the shape of their ears or by other physical characteristics. Since then, experts have agreed on the existence of two African subspecies or species: the smaller round-eared or *forest elephant* (Loxodonta cyclotis), which lives in West Africa and the forests of the Congo, and the large-eared or *bush elephant* (Loxodonta africana). The latter is found in South and East Africa as far north as Ethiopia and Somaliland. The forest elephant has five toes on its fore-feet and four behind, whereas the bush elephant is generally assumed to have four in front and three behind. The forest elephant never moves the upper edges of its ears, unlike the bush elephant, whose ears attain a far greater size, are roughly triangular, and taper to a point at the lower lobe. The forest elephant stands 7 to 8 feet at the shoulder, seldom more, whereas bush elephants usually carry their heads higher and are taller: cows on average 9 feet, bulls $10\frac{1}{2}$ feet and in exceptional cases $11\frac{1}{2}$ to 12 feet. Above all, though, forest elephants have thinner, downward-projecting tusks, whereas the bush elephant's more powerful tusks point forwards past the trunk and usually terminate in an upward curve. The ivory of the forest elephant is harder.

However, the two types are not easily distinguishable unless one contrasts a forest elephant from the West Coast with a bush

Do pygmy elephants exist?

elephant from the East Coast. Many transitional varieties occur where the two types overlap. Some of the specimens I saw at Gangala na Bodio could have been identified as bush elephants by their ears or forest elephants by their number of toes, and the elephants in Queen Elizabeth Park beside Lake Edward in Uganda have tusks which are quite reminiscent of forest elephants. The two types or species readily interbreed.

What of the celebrated pygmy elephant? The German zoologist Noack introduced it into zoology in 1906 as a new species designated Elephas africanus pumilio. He based his description on a single live specimen. This had tusks consistent with a putative age of six but was only just under 6 feet tall, a height normally attained by young elephants after only eighteen months. The animal survived in New York for nine years and was only $6\frac{1}{2}$ feet tall when it died. Since then, full-grown elephants less than $6\frac{1}{2}$ feet in height have been regarded as pygmy elephants. Most of those displayed as such in circuses and animal shows are undoubtedly young animals, but long-tusked specimens of similar size have repeatedly been sighted in the wild, notably in West Africa. Major Powell-Cotton saw a cow in calf with a shoulder-height of only 6 feet. The elephant-training centre once had two particularly small beasts which were only 4 feet 3 inches tall when captured but already had tusks 2 feet 3 inches long, which would have made them 12-14 years old. The animals grew to a height of only 5 feet 2 inches in the succeeding ten years, and their tusks to a length of 40 inches. Unfortunately, they were then sold. Pygmy elephants of this type are far from being confined to particular areas, but they never occur in large herds. We may therefore conclude that they do not belong to a special breed, far less species, and are quite as much isolated examples in a population of normal size as the exceptionally huge elephants that are also found.

It was extremely difficult and dangerous for the Africans to hunt their continent's heaviest animal while they still had no effective fire-arms. Meat was much more easily obtainable from countless other species of game. The real lure of elephant-hunting consisted in the tusks, which had been traded in from time immemorial and fetched high prices.

Accordingly, herds of elephants were ringed with artificial walls of flame, driven into traps, showered with poisoned arrows. Before the days of railroads, ivory had to be carried from Uganda to the coast by porters, a ninety-day march. It was the only commodity which could repay such a costly form of transportation. The explorer Sir Samuel Baker estimated that traders in ivory made a profit of at least 1,500%, and often earned as much as 2,000% himself, yet the hunters who supplied it were equally content. It is probable that most of the tusks were taken from elephants which had died of natural causes. In the colonial era, hunting and ivory-trading became subject to official approval and were supervised by game departments. According to Noel Simon, however, Kenya still loses revenue well in excess of £100,000 per annum through poaching and well-organized smuggling.

Even today, and particularly in East Africa, many elephants have to be shot because more and more land is required for the growing human population. In 1963, for instance, the Tanzanian game authorities had 3,247 elephants shot by their own employees and issued 393 permits to other hunters. They obtained 16 shillings per pound for ivory at auction, or a total of 103,540 East African pounds for approximately 58 tons of ivory. But, whereas between 1850 and 1860 the East African elephants' tusks that reached the Zanzibar market weighed 50-100 lb. and still averaged 40 lb. in Rhodesia at the end of the last century, the tusks of 31,966 elephants shot in Uganda between 1927 and 1958 by employees of the Game Department averaged approximately 26 lb. Less than 2 per cent of the tusks weighed more than 50 lb. In 1929 elephants were still to be found in approximately 70 per cent of Uganda as opposed to 17 per cent thirty years later, and their spread has probably narrowed considerably since then. Belgian game department officials told me that it was quite common for up to 2,000 elephants to assemble at certain times of the year in the North-East Congo, though the reason was unknown. One of Livingstone's travelling-

A declining industry

companions once saw 800 elephants together beside the Zambezi. Today, spectacles of this kind are seldom to be seen in Africa.

How does the general decline of the elephant fit in with newspaper reports which tell of thousands of elephants being killed to preserve vegetation? After the severe droughts of 1960 and 1961 I found trees in the rivers of Tsavo National Park in Kenya which had been uprooted or barked by elephants. Even the enormous baobab-trees, with their fat-bellied, sap-filled stems, had been dug up and felled. In the Murchison Falls National Park of Uganda, there are very many square miles of land where scattered trees are constantly being killed off by elephants which strip them of bark. Elephants have recently adopted similar behaviour in the Congo's Parc National Albert, in South Africa's Kruger National Park, and elsewhere. E. Davison observed in Rhodesia's Wankie National Park that the acacia-trees were only barked by a few individual cows in each herd, not by all and never by large bulls.

In June and September 1962, counts were made from the air of all the elephants in Tsavo National Park, which covers 8,000 square miles. This elephant census, which was fairly accurate because of the size of the animals and the almost treeless terrain, yielded figures of 6,825 for June and 10,799 for September, to which should be added 4,804 head sighted in adjoining areas. There were 1,007 herds and 128 lone elephants. Some 300 of the herds consisted of 2-5 elephants, and a similar proportion of 6-10, but ten concentrations of more than a hundred head were sighted. The three largest comprised 191, 289, and 700 animals.

The Tsavo National Park's 8,000 square miles make it roughly half the size of Switzerland, so 10,000 elephants are not, on the face of it, a very large population. The elephant-density was 0.34 per square kilometre in June and 0.54 in September. These figures are small in comparison with those prevailing in other parts of Africa: 1.1 elephants per square kilometre in the Aberdare Mountains of Kenya; 1.7 in the Ruindi-Rutchuru plains in the Congo's Parc National Albert; 1.72 in Uganda's Queen Elizabeth Park; and 1.8 in the Murchison Falls National Park, also in Uganda. And yet there are too many elephants, when one compares the Tsavo area with other national parks and takes account of the sparse vegetation that grows in such drought-ridden regions. Almost all the animals were found at a distance of 14 miles from the nearest permanent water. Although a more recent tally made in 1966 indicated that Tsavo's elephant population had climbed to 20,500 head, I drove round the district in January 1966 and saw a total – from the car – of two. Tsavo National Park has only one river which carries water throughout the year, namely, the Galana. Elephants so badly ravaged the riverside vegetation during the two years of drought that in 1960 about 300 out of the 780 or so rhinos starved to death.

Once bush-country, with tall tropical river-forests bordering its rivers, Tsavo is progressively turning into savannah. The elephants will probably suffer least as a result because they survive well in pure grassland. Worse hit will be the hook-lipped rhinos, which browse principally on twigs and undergrowth and do not travel as far as elephants to avoid drought. Moreover, in contrast to many trees and bushes, whose roots penetrate the ground far more deeply, the grass of the savannah becomes completely parched during the dry months and yields little moisture when grazed – indeed, disappears entirely in the wake of frequent bush-fires.

The main reason for the damage inflicted by elephants in Kenya's Tsavo National Park is the artificial creation of watering-places. This is the conclusion arrived at by, among others, Sylvia K. Sikes, who has made a study of elephant problems in East Africa. In her opinion, the trouble started when water had to be provided for the construction of the Mombasa-Nairobi railway. This attracted large and water-dependent animals, especially elephants, adult males of which species require 20-28 gallons of water daily. The elephants continued to loiter near the railway line during the dry season because human habitations had sprung up there. They came to drink at night and quietly slipped away before daybreak. The position became even worse when the administrators of Kenya's national parks installed permanent artificial drinking-

Man-made waterholes misguided

places in Tsavo. Elephants now remained all year round in areas where the vegetation could not provide them with sufficient nourishment. The inevitable result was that they devastated the surrounding country. Sylvia Sikes concludes that a determined effort must be made to combat bush-fires and that man-made watering-places must be shut down entirely during the dry season so as to drive elephants out of the park. What causes local overcrowding elsewhere? Unlike many other African animals, ele-

Page 269:
Gnus have multiplied steadily in the Serengeti Plain since 1957-8, when I and my son Michael made the first accurate count from the air. They now number four times as many, or about 1.3 million adult animals. Here, gnu herds are migrating across the savannah, their vast numbers covering it for as far as the eye can see.
Repeated attempts have been made in South Africa and Rhodesia to tame the eland and convert it into a domestic animal. Herds of domesticated eland are, in fact, kept on one or two farms today. In 1896 the South Ukrainian estate-owner Friedrich von Falz-Fein acquired some eland whose descendants still live in the Ukraine today and are grazed on the steppe like cows by mounted herdsmen, some of them being regularly milked. This is a bull eland with Lake Victoria in the background.

Page 270:
African wild dogs have been persecuted and shot for a century-and-a-half because of a total failure to appreciate their character and role in Nature. Our conception of them has changed remarkably in recent years as a result of breeding in zoos and months of observation in the plains of East Africa.

Page 271, above:
Three greater kudu bulls are here seen coming to drink together in South Africa's Kruger National Park. In the background, a female impala. The greater kudu is very common in some parts of South Africa. A full-grown kudu bull leapt on to the road between Johannesburg and Pietersburg at three o'clock one afternoon and landed on top of a car containing a European and an African. Since the car was travelling at 70 m.p.h., the animal tore the roof off and ended up dead on the back seat. The European was severely injured and taken to hospital; the African, who was wearing a safety-belt, escaped virtually unhurt. This may seem a rather distressing caption to associate with such handsome creatures, but annual figures of wild life killed elsewhere by cars make equally distressing reading: 660 stags and 44,000 deer in the Federal Republic of Germany; 1,200 elk in Sweden; no less than 77,000 deer in the United States. In Germany such accidents also take a considerable toll of human life, merely because people are reluctant to spend a little money on game barriers when constructing highways.

Page 271, below:
Zebra families stick together. Stallions defend new-born foals from attack by hyenas, and uncles and aunts usually rally round as well. White and maneless zebras sometimes occur, also checked ones. No one zebra has precisely the same stripes as another, which is why the animals can be identified photographically like human beings by their fingerprints. These are Damara (Chapman's) zebras from South-West Africa.

phants are persevering travellers over long distances. They are also being doggedly hunted and exterminated in most parts of Africa today. Consequently, it is only natural that they should turn off into the few protected areas and overcrowd them to such an extent that they become a menace to the country side. It may also be that they are cut off from areas in which they earlier satisfied their needs for mineral salts. Tree-bark and twigs contain not only moisture but, for example, 3.4-5.68 per cent calcium carbonate, whereas the comparable figure for grass is only 0.18-0.33 per cent.

Where these problems are concerned, we are still groping in the dark. Although human beings have pursued and admired elephants for centuries, and although so many big-game hunters have written highly-coloured books about them, we know precious little about their needs or way of life. That is why the Kenyan authorities were wise not to embark immediately on the slaughter of four or five thousand elephants, as the newspapers recommended. Instead, they began by summoning a team of scientists to study the animals' needs, an investigation which the Ford Foundation has subsidized to the tune of £75,000. It is, in fact, possible that the destruction of trees should also be attributed to man-made grass-fires, which are steadily increasing in frequency. They too destroy trees and create the open grassland that becomes transformed in certain areas into thorn-bush and other forms of scrub, a process which is already under way in some parts of Tsavo National Park.

It is shaming to reflect that when we Europeans ruled Africa as colonialists we made little effort to study the living habits of large animals which are so important to balance of nature, climate and human survival in these torrid regions. Only in recent years, and,

Page 272:
The greater kudu has been rare in East Africa ever since the great rinderpest outbreak which occurred towards the end of the last century but is still quite common in South-West Africa, Rhodesia, Mozambique, the Kruger Park, and down as far as Zululand. Greater kudu are among the largest and most handsome antelopes. They will readily clear an eight-foot fence.

curiously enough, only since the African countries gained their independence, have biologists really tackled the many problems posed by this giant among animals. The most impressive study-centre is the Serengeti Research Institute at Seronera in Serengeti National Park, set up by John Owen, the highly successful national parks director. 10-18 zoologists, botanists, geologists and experts on animal behaviour from a wide variety of countries are permanently engaged in research there for tours of not less than three years. So we are now gradually gaining a picture of the African countryside and its wild life under conditions of aridity. The scientists' living accommodation and the Michael Grzimek Memorial Laboratory were built by the Fritz Thyssen Foundation.

It was only after weeks of endeavour that I sighted my first wild elephant in the forests of the Ivory Coast in West Africa. We had installed ourselves on a large banana plantation owned by an absentee landlord. The plantation was visited every night by a herd of wild elephants, shy but intrusive creatures which used to make their way along the bed of a stream running through the middle of the banana-groves. A small bridge of logs had been built to enable cars to negotiate it with greater ease, but this did not suit the elephants at all. Instead of climbing over or making a detour round the obstacle when proceeding along the stream-bed, they regularly picked up the logs and laid them to one side.

The elephants took strong exception to many of the man-made innovations in their habitat. Low sign-posts they often covered with piles of branches or tore out and dragged for short distances. They even deposited layers of branches on newly laid stretches of road. Telegraph-poles which had been laboriously dug into the ground were sometimes uprooted in dozens by nocturnal raiders and laid flat one after another.

Film sequences which show herds of elephants attacking native villages, trampling huts and crushing human beings to death are a complete travesty of the truth. Incidents like these do not occur in real life. Having no natural enemies, however, elephants do sometimes hang around huts for hours and may in exceptional cases strip part of a roof or push down a wall if there is something inside that tickles their appetite or excites their curiosity. In Ezo, in the Southern Sudan, three persistent intruders were easily dispersed by loud waltz music when a visiting agricultural inspector turned his car radio up full.

One recent phenomenon has been the emergence of 'tourist elephants', lone beasts which are becoming more and more of a pest at tourist centres because visitors have ignored the ban on feeding them. I have described the life-story and sad demise of Lord Mayor at Paraa Lodge in Murchison Falls Park in my book *Rhinos belong to everybody*. Last year I was repeatedly followed by a half-grown elephant named Charly as I left the restaurant building at night, bound for my room in one of the bungalows. He examined me so closely that I decided to hug the walls and cross any open spaces at speed.

One evening not long afterwards the young bull posted himself outside the dining-room windows and watched the guests at their evening meal. He behaved like a small child staring into the window of a sweet-shop – in fact his head was so close to the panes that his short tusks broke three of them. Not content with that, he stuck his trunk through the open window and groped for food. A few days later the manager of the restaurant had to repel the importunate animal by belabouring it with a stick. Another local elephant, this time female, earned the nickname Dustbin Nelly because she and her calf Billy were so fond of inspecting the trash-cans. When one of the tourists tried to take a close-up of Billy on the veranda, Nelly went for him. The game warden just managed to grab the visitor by the arm and push him inside. Nelly pulled up 6 feet short of the game warden and then retired.

Another time, Billy took an enthusiastic interest in the wire ropes supporting the radio mast. He tugged at them so hard that the upper section of the mast toppled and fell across his mother's back. Both animals galloped off into the bush. Billy also removed the notice in front of the hotel warning visitors to beware of elephants.

Severed trunks

Billy twice tried to push cars out of the car park. He met an untimely death at the age of two-and-a-half, probably because of the indigestible things he had purloined from the garbage cans. The dung of elephants which make a habit of raiding dustbins has often been found to contain plastic bags and wrappers.

Probably because of their playful disposition, elephants quite often get caught in the steel wire nooses set for other game by poachers. Once the noose has drawn tight about the trunk it cuts deeper and deeper into the hide, inflicting an annular wound which soon starts to fester. I myself have seen two such victims. A game warden at Acholi in Uganda also came across an elephant which had caught its trunk in a wire noose. The elephant had vanished by the time he returned with an armed companion, but three feet of its trunk still hung from the noose. An almost trunkless elephant was sighted in Malawi. It had to graze in a kneeling position but seemed quite well nourished.

There are few ways of stopping an elephant from going where it wants, short of shooting it. A few years ago, a game fence was erected on the Sanyati River, about 65 miles downstream from the well-known Kariba Dam in Zambia. It was designed to prevent wild animals from leaving the Zambezi Valley and invading the densely populated areas near by. The new fence, which ran straight through the forest, was often damaged and breached by elephants. For that reason, broad strips of land were cleared on either side and watchmen posted to drive off the game with warning shots. In general, the scheme worked. The fence consisted of eight wires, each of them with a breaking-strain of one ton. The hardwood posts were sunk to a depth of three feet, so firmly that they might have been cemented in. The wire strands whipped back painfully if they were broken.

One day, despite all these precautions, two elephants crossed the wire. Having vainly tried to breach it by force, they flew into a rage and broke down some forty trees in the vicinity. They then hauled six trunks about ten inches in diameter to the wire and deposited them on top. The wires sagged so low under their weight

A lingering death

that the elephants were able to scramble over the fence. A week later they made similar use of a far heavier tree to get back again.

Anyone who wants to learn how to build elephant-proof fences would be best advised to visit South Africa's Addo National Park, which covers an area of only 26 square miles and is situated near Port Elizabeth at the southern tip of the African continent. This small patch of bush in the middle of an increasingly populous area was declared a national park in 1931 and became a sanctuary for the dozen-odd South African elephants which still survived. After an initially successful period during which their numbers increased to 25, the position deteriorated and by 1949 had become quite untenable. No less than 30 farms worked by 150 Europeans and nearly 300 Africans had been established on the boundaries of the park, which the elephants crossed almost every night. Fences were trampled, water-pipes torn out of the ground, dams destroyed, gardens and plantations laid waste, and one or two domestic animals killed. Moreover, people became afraid of the huge beasts, which would suddenly turn up without warning in the middle of the night. The farmers, really in danger from time to time, shot at the elephants, and wounded animals went on to threaten other people, notably game wardens. One night, five elephants crossed the railway line. Just as they were returning in the dawn light as usual, a goods train bore down on them. Four of the beasts got across the track in time, but the fifth, a cow, hesitated. The train screeched to a halt, but not before it had fatally injured the she-elephant. This occurred precisely a fortnight after the leader of the herd had been killed in a similar incident.

Another bull repeatedly attacked farm-workers, who were forced to run for their lives and scramble up the steel framework of wind-pumps. One day the bull attacked a train, got the worst of it and was quite badly knocked about. The park had to be closed for a while on his account. A year later he again met an approaching train. Obviously remembering his unpleasant experience, he turned and made for the park, but the train caught him and broke one of his hind legs. Screaming with pain, he hobbled for twenty-

The fence that held

five yards on three legs and then collapsed. His death-cries could be heard all night, but it was impossible to put him out of his misery. The other elephants had surrounded him, so he had to die slowly where he lay. That reduced the elephants' numbers to seventeen, of which only three were bulls of breeding age.

For all these reasons, it was decided to fence the park off. Experiments were made during the 1940's with an electric fence, but the elephants broke out repeatedly and wrought havoc in the neighbourhood. It was a long time before the national parks raised enough money to build a fence of sufficient strength. Work progressed from 1951 to 1957. The fence consists of sections of old tram and railway track driven 6 feet into the ground and projecting 8 feet above it. These are placed at intervals of 25 feet, and between them, likewise driven into the ground and 8 feet apart, stand hardwood stakes. Suspended from, and linking, the strands of wire are logs which have been studded with sharp protruding nails to deter the elephants from grasping them with their trunks. The strands themselves are steel lift and mine-cables under great tension.

This fence has readily withstood the full impact of a charging she-elephant. In all, 10 square miles of impenetrable thorn-bush have been enclosed in this expensive fashion, and the elephants have multiplied satisfactorily in the interim. A similar but far shorter fence can be seen in the forest at one end of Manyara National Park in Tanzania, between the lake and the steep wall of the Rift Valley.

I have only once been attacked in a car, and that was by a cow in Queen Elizabeth Park. My son and I had got out and set up our camera-tripod, intending to film her calf. We jumped back into the car when she charged, but she stopped just short of the vehicle and refrained from touching it when the engine did not start. Elephants seldom attack cars, but two drivers who were sleeping beside the Lodge in Murchison Park were suitably startled when an elephant's tusk appeared through the open rear window of their car.

Don't blow your horn

On another noteworthy occasion, three American professors and their students were driving through the Congo's Parc National Albert when they were suddenly and for no apparent reason attacked by two bulls. Professor Gevers managed to film one of the attackers through the car window before it mangled the engine, skewered the body-work with its tusks and turned the vehicle over, breaking the professor's legs. It then ran off, but abruptly fell dead after a short distance. The cause of death could not be determined. Professor Gevers was given one of the animal's tusks as a souvenir of his adventure.

In 1965, a group of elephants held up the traffic for fifteen minutes on a road in Kruger National Park, South Africa. One of the cars, a Volkswagen, tooted impatiently, whereupon the bull elephant which had been standing in the road covering the cows and calves as they slowly crossed, turned and spread his ears. Then he bellowed and bore down on the little car. Having hoisted it into the air by inserting his trunk and tusks under the front axle, he buckled the front of the vehicle with a tusk, lifted it twice more, and finally pushed it five yards off the road into the bush. The occupants, a married couple named Bauer, escaped injury but were looking rather green about the gills when they extricated themselves. 'Mr Bauer, who is a garage-owner, intends to repair the car himself,' reported the *Cape Times*.

Artists generally depict African elephants charging with trunks raised and ears spread, but the animals seem more often to behave in this manner when they wish to investigate another creature by means of scent and hearing. I myself have never seen elephants charging in earnest with their trunks raised, even though I have photographed them at close range. Other people have confirmed this. An elephant which charges an enemy normally rolls its trunk back against its chest, puts its head down, and looks progressively 'smaller' as it nears its target. My experience was also borne out in Rhodesia by a Mr G. Gilett, whose car was charged head-on by a female elephant and had two holes drilled beneath its radiator by her tusks. Mr Gilett, who was unhurt, reported later that 'she laid

A close call on a bicycle

her ears flat against her neck, and pushed the car back thirty yards before I could stop her'.

In 1948, two years after accompanying me to the Congo, a former camera-man of mine was almost killed while filming beside the Nyamugasani river in Queen Elizabeth Park, Uganda. Having got out of his car in defiance of regulations, he was unexpectedly charged by an elephant. It dragged him out of the bush in which he took refuge and tossed him into the air three times, but his only injuries were a tusk-wound in the leg and a broken ankle. During the same year two Africans were killed just outside the park boundaries by an elephant which was later shot. Incidents of this kind are attributable to wounds inflicted on elephants by poachers and to the ineptitude of some of the Europeans who are issued with game licences. As usual, it is the innocent who suffer in the end.

In July 1959, an African roads superintendent was cycling through Murchison Falls Park to the hospital at Masindi because his pregnant wife had been taken ill there. In his haste and preoccupation, he suddenly found himself in the middle of a herd of elephants some miles down the road. 'There were elephants to the left of me, to the right of me, in front of me and behind me – elephants everywhere. I got off my bicycle and stood there for a while without moving. My presence didn't seem to have startled them, although they eyed me closely. After a short time the animals began to move off and my fear subsided. Then, suddenly, one mother-elephant attacked me from the flank and another from the front. I threw down my bicycle and ran faster than I had ever run before.

'The she-elephant took no notice of the bicycle, so I threw her my big overcoat. That didn't interest her either, so I jettisoned a shoe. She didn't want the shoe, so I picked up a stick. The elephant grabbed the other end of the stick, and we continued to run like that. I felt as if I was flying, but I tired very quickly. I lost ground whereas the elephant ran faster and faster. Just as I felt her trunk almost touching me I tripped and fell between her forelegs. The cow lowered her massive head and gored the ground with her tusks, pinning me between them.

'I somehow managed, in my desperation, to slip out of my jacket and thrust it right into the mouth of the furious beast with my left hand. She kicked me with her foot and ran off. I lay there on the ground, half stunned, but managed to crawl another two hundred yards or so on my knees. When I had fully recovered consciousness I went back to the road-workers' camp. From there someone took me by bicycle to Masindi Hospital, where I received treatment.'

During the exceptionally dry summer of 1951, even elephants died of thirst in Northern Kenya. They competed for water-holes with the Somali natives and attacked women carrying water-containers. Some of the water-holes in this area are sunk deep into the ground, and the natives have to carry the water to the surface by way of ladders running up the sides of these deep shafts. They did not dare go home until their clothes had completely dried out because the elephants were so thirsty that they would have been attracted by the moisture and followed them.

No one has yet found a satisfactory explanation of the fact that elephants often use branches to cover human beings whom they have killed or whom they come upon dead or apparently dead. In 1954 a blind old Turkana woman from the north of Kenya lost her way and was overtaken by darkness. She found a tree whose spreading branches almost brushed the ground, crawled beneath them, and fell asleep. During the night she was disturbed by the trumpeting of an elephant. The animal groped for her with its trunk and must have gained the impression that she was dead. Breaking off branches from the tree beneath which she lay, as well as from other trees near by, it carefully deposited them on top of her. When found next morning by a search party she was five feet deep in leafy branches and totally incapable of extricating herself unaided.

A farmer near Ruhengeri in the Congo attempted to drive an elephant off his pyrethrum plantation by firing at it but was attacked by the wounded animal and tossed high into the air. He

Elephants bury human victims

broke several ribs as he hit the ground, and passed out, but awoke some time later to find himself buried beneath a pile of branches. similarly, a game warden who had been killed by an elephant in the Kilwa district near Makumba, Tanzania, was later found covered with earth. His killer had topped the mound with layers of brushwood.

Dr Wolf Dietrich Kühme, who has done research at the Michael Grzimek Laboratory in Serengeti, once lay down near a pile of refuse six feet from the African elephant enclosure in the Opel Zoo at five o'clock one morning. The master-bull fished for branches and pelted Kühme's recumbent figure with dung, twigs, and gravel. The animal fired about fifteen salvoes in the half-hour that elapsed before its attention was distracted by the keeper. When Dr Kühme stood up, the shape of his body was clearly discernible as a blank space in the midst of all the mess.

It has long been known that tame elephants become familiar with certain individuals – zoo keepers, for instance, or animal trainers in a circus. Most of these observations relate to Asiatic elephants, but as far as I could ascertain the tame African elephants at Gangala na Bodio behaved no differently. It struck me and my son Michael that a group of these working-elephants, which were accommodated in the Ituri Forest near the Okapi trapping centre, treated us with great diffidence. Michael accordingly blacked his face, arms and legs and put on an elephant-rider's uniform. Animals which had always backed away nervously before now allowed him to mount and ride them without hesitation and obeyed his commands. An account of this, complete with photographs, appears in my book *No room for wild animals*.

Many features of an elephant's behaviour towards man become easier to understand if we take a closer look at how elephants behave towards their own and other species. Apart from modern man with his sophisticated weapons, their environment contains no enemies or rivals. Probably because of this, the big animals are tolerant towards other living creatures. Waterbuck, impala, Cape buffalo and small antelopes can be seen grazing unconcernedly only a few yards from the grey colossi. All other animals, even rhinos, hippos and lions, avoid full-grown elephants or retreat before them, e.g. on narrow paths enclosed by undergrowth or steep banks. B. Nicholson once watched a small herd of elephants, only one of which had tusks, on the banks of the Kilombero in Tanzania. They were led by a cow, also tuskless. Suddenly, the latter gave a start and quickly retreated with her head in the defensive position. The next moment a rhinoceros emerged from a dip in the ground and made straight for the cow, snorting angrily. Having come within four or five yards, it lost heart and veered away. Three times it charged the group of elephants, and each time it turned aside before reaching them. It finally made off with its little tail erect. The elephants had merely defended themselves the whole time without attacking. On the other hand, a fierce battle was observed in Kruger Park between an elephant and a rhinoceros. The rhino ended up dead with four tusk-wounds in its body.

Elephants take little notice of crocodiles, judging by the casual way they walk in and out of water. While in a motor-boat, watching a group of elephants drinking on the banks of the Nile in Murchison Falls Park, Colonel Radford saw one of them start back in alarm. Suspended from the elephant's trunk, into which it had sunk its teeth, was a 5-foot crocodile. The elephant shook it off and departed hastily.

Much the same goes for lions. A. Schiess once saw a big bull elephant in the Etosha Pan in South-West Africa make a series of short but furious charges in the direction of a large marula-tree. Changing its mode of procedure, it circled the tree and charged again, and this time the long grass beneath the tree moved. 'Then it happened. The big animal, either shamed or infuriated by its own cowardice, suddenly charged straight at the tree. The result was a regular explosion. As though propelled by a charge of dynamite, four lions shot into the air, only to vanish once more into the long grass. The exception was a young lion which stood its ground and eyed the giant. The elephant halted abruptly with a comical movement of the sort one sees in Disney films. For a few seconds, elephant and lion stared motionlessly at one another, each poised to run away at the slightest movement on the other's part. In the end Jumbo lost his nerve and retired, trumpeting shrilly, whereupon the lion calmly returned to the tree. To judge by the movements in the grass, the others had also calmed down. The elephant stood there for a while, staring at the tree, then departed in the direction of the pool. A thought must have occurred to him as he passed a clump of bushes. Keeping under cover of the bushes he again approached the tree, more cautiously this time. We lost sight of him for a while. Then his huge head suddenly appeared above the bushes, trunk high in the air. The trunk moved slowly until it pointed straight at the tree like a snake about to strike. The lions kept completely still. Then the elephant turned, gave one last look, and went down to the water to drink.'

A game warden in the Wankie game reserve once had to shoot an ill-tempered old bull elephant with a very severe wound in its leg. He found that the bull had been standing with one foot on a wart-hog which had evidently been killed by a cheetah. The elephant had unaccountably driven the cheetah off its prey and remained there with its foot on the carcase. The game warden dragged the wart-hog some distance from the dead elephant and left it there. The cheetah returned soon afterwards and began its meal.

Elephants somtimes take it into their huge heads to do unaccountable things. In September 1957, thirteen elephants were standing beside a small marshy water-hole in one of the empty craters in Queen Elizabeth Park. About twenty yards from them waited two full-grown Cape buffalo with a sizeable calf. One of the elephants seemed to resent this. After making a series of mock attacks, it seized the buffalo calf with its trunk and made as if to dash it to the ground. Eventually, however, it trampled slowly about on the luckless and badly injured young animal while the mother-buffalo launched repeated but unsuccessful attacks on it. Having tried to finish the job with its tusks, the elephant departed. The mother-buffalo went over to her calf, which was still moving. It even rose on its forelegs, but its hind quarters were so badly smashed that it collapsed when it tried to follow her. The cow seemed to realize that it was useless to linger, and slowly rejoined the other buffalo.

African labourers at Letaba Camp in Kruger National Park were not infrequently disturbed by wild animals, so the game warden relaxed the usual ban on dogs and allowed them to keep one. This dog put elephants to flight on numerous occasions. The elephants began by trying to kill the dog, which used to bark at them and nip their legs from behind, but failed to catch it. The local tuskers got to know it so well that they never ventured within a hundred yards of the camp. The dog took no notice of them if they were far enough away.

This may have been because the dog was a new and unfamiliar animal. Like horses, wild elephants are exceptionally suspicious and nervous of the unknown, though it is untrue – judging by my own experiments – that elephants are scared of mice. Nor do mice run up elephant's trunks and asphyxiate them, as natives in many parts of Africa and even some white game wardens allege. Not only could an elephant blow a mouse out of its trunk without difficulty, but it is capable of breathing through its mouth. Zoo elephants readily put the wide-open ends of their trunks close to mice in order to sniff them. On the other hand, when I introduced rabbits and dachshunds into circus elephants' stables or zoo elephants' enclosures the great beasts backed away nervously and pelted them with sand and stones from a distance by kicking at the ground with their forefeet.

Is it possible to outrun a charging elephant? An African elephant walks about as fast as a man, covering 3 to 4 miles in an hour, but it can almost double this pace and maintain it for hours on end. When elephants flee in panic or charge, they attain speeds of up to 20 miles an hour. This makes them faster than a human sprinter but slower than saddle-horses, antelope, and most feline predators. They cannot, however, maintain their maximum speed for more than a few hundred yards. Bush-country obstructs them less than it does man or smaller animals because they can simply burst

Elephant sleeping habits

through the undergrowth. They cannot gallop or jump, and vertical walls of moderate height represent insurmountable obstacles to them. At Nuremberg Zoo elephants have for years been living behind a ditch which is 5 feet 6 inches wide at the top and only 4 feet 6 inches deep at its lowest point. Most ditches are over 6 feet 6 inches wide, but only to prevent elephants from falling in and becoming wedged. On the other hand, the heavy beasts are excellent at scaling steep slopes and do, as we have already said, climb high into the mountains. We had difficulty in following some of the elephant tracks in the Congo because they usually led straight upwards. Elephants have been known to cover more than 50 miles in a single night.

Elephants were long reputed to sleep standing up and never to lie down. Professor Hediger demonstrated in the case of Indian circus elephants that they do, in fact, lie down to sleep, preferably just before midnight. Old elephants lie down for only two to three hours, younger ones for longer. I myself have never seen a wild elephant lying down or sleeping in a recumbent position, but I have come upon places in the jungles of the Ivory Coast where elephants had been sleeping until a short while before, because every contour of their bodies was clearly imprinted in the soil. On cautious nocturnal tours of inspection at Gangala na Bodio elephant station, I found all the tethered elephants standing before midnight and twelve out of sixteen lying down after midnight. I was even able to take flashlight photographs of them. It seems that zoo elephants rise to their feet very swiftly if someone so much as inserts a key in the lock at an unusual hour, which may explain their not having been observed in a recumbent position for so long. On the other hand, recumbent elephants do sleep very soundly for at least part of the time. Our elephant keeper was once accidentally locked up in the elephant house and could not leave by the main door. He climbed over several sleeping elephants in order to reach another exit, but none of the animals moved or got up. They were, admittedly, well acquainted with him. Early one morning in 1963, one of the labourers employed at Murchison Falls National Park in Uganda saw what he thought was a dead elephant lying near Paraa Lodge. Anxious to earn a fee for securing the ivory, he left a companion to guard the carcase and quickly ran off to fetch an axe. He returned with the axe and approached the elephant with two other men. Just as a precaution, he threw a stone at the animal. 'A very surprised and angry elephant got to its feet and all concerned fled, fortunately in different directions.'

The elephants at Gangala na Bodio did not allow their slumbers to be disturbed by the nudges and grunts of their own kind, but one flash from my flash-gun brought them to their feet at once. Some of them made regular pillows for themselves out of the branches that were brought them as fodder. The same thing can be observed with zoo elephants. Elephants asleep on their feet – especially bush elephants, whose tusks are more downwards-inclined – often prop their tusks on the ground. The trunk touches the ground too.

Elephants can swim but seldom do so despite their fondness for bathing. They can often be seen swimming the Kazuga inlet near Mweya Lodge in Queen Elizabeth Park. The fast-flowing Victoria Nile in Murchison Falls National Park, on the other hand, forms a genuine frontier between the two elephant populations on either bank. Only a few lone animals have ever been seen swimming the river. In 1965, the game warden approached a swimming bull elephant in his motor-boat. 'The elephant was rather disconcerted by the boat's approach but soon swam on. It is impossible to describe the method of swimming. The animal submerges completely at times. Then the trunk reappears, followed soon afterwards by the head. The whole thing was reminiscent of a human swimmer doing the crawl.'

Another elephant was spotted swimming in the same direction halfway between Paraa and Namsika. A much older bull with drooping ears, this animal seemed to find the act of swimming a great physical strain. It had considerable difficulty in climbing out of the water and on to the bank.

Elephants trumpet shrilly when agitated. This may occur when

Elephants help each other

charging or in response to a sudden shock, or when a single animal has strayed and is feeling lonely. Wounded elephants are said by hunters to emit genuine deep-throated roars, but I have never heard this myself. Many theories have been advanced about the purring or rumbling sounds made by elephants. It was, in fact, assumed until recently that the animals produced these sounds with their stomach or intestines. What militates against this assumption is that they can stop them at will, e.g. when stalked by human beings, and that sleeping elephants have never been heard to make them. In other words, the sounds are subject to the will, whereas stomach and intestines have muscles which operate independently and cannot be controlled by the brain. The animals are said to make them almost invariably when in groups, not when they are by themselves. This indicates that they represent a means of transmitting mood, an assurance that the elephant's companions are still there, and an intraspecific recognition signal for use when other elephants are approaching. The elephant expert I. O. Buss, who had spent a long time in Uganda investigating this problem, subsequently solved it at Basle Zoo. There, African and Indian elephants could be induced to 'rumble' by the long-familiar voice of their keeper. As soon as one of them started, the keeper and Buss were hemmed in by the other animals, which pressed round them. By placing one hand on the throat of a 'rumbling' elephant and the other on the end of its trunk, Buss could distinctly perceive the vibrations in the animal's throat and the concomitant jets of air from its trunk. It remains to be seen how much of a role this rumbling plays in sexual relationships and in the ranking order within a herd.

Elephants have often been seen trying to help and support a wounded member of the herd. One such instance has been described by W. D. Nicholson, Elephant Control Officer in Tanzania. Nicholson had smashed the shoulder of a bull elephant near the Ruvuma river in the district of Tunduru. Two cows which had already started to run off in the opposite direction promptly returned when they heard the bull's cries for help. Stationing them-

They cover up their dead

selves one on each flank, they tried to half-carry, half-drag him away. Nicholson was worried in case they should succeed, so he ran along beside the trio and shouted in an attempt to drive the cows away. At the sound of his voice, one of them turned and went for him. He had no alternative but to shoot her and give the bull the coup de grace. The shots had meanwhile scared the other cow away. When she saw the wounded bull fall to the ground she seemed to realize what had happened and walked sadly away, pausing every few yards to listen and look round.

There are many recorded instances in which elephants have tried to lift animals that have been shot and killed. On numerous occasions, whole herds of trumpeting and screaming elephants have returned to the scene and tried to raise one of their fallen companions. Many hunters have interpreted this as a concerted attack.

The Zambian game warden Patulani Ng'uni was once instructed to drive away four female elephants which had persistently devastated gardens and plantations. He shot the lead-elephant, whereupon one of the others pursued him for half a mile before returning to its dead companion. When Ng'uni himself returned to the scene he found the survivors gathered round the carcase. They were trying to lift it, using their trunks to grasp the head, tusks, legs, and body, but part of the carcase always remained on the ground. One of the dead elephant's tusks broke off in the process and was thrown for a distance of five yards. The would-be rescuers remained at the scene for about four hours, breaking down trees. Next morning Ng'uni found the dead elephant completely covered over with mopani branches. Another game warden, also in Zambia, watched elephants beside the Katete river vainly trying to lift the head and forelegs of a shot companion for one-and-a-half hours. Nicholson has established that bulls will try to help a fallen bull and that cows have done the same for members of both sexes. He has never, curiously enough, seen a bull make the slightest attempt to help a fallen cow.

David Sheldrick, long-time game warden at Tsavo National

Ear-movements and ranking order

Park in the south of Kenya, states that elephants very often strip tusks from the carcases of their dead and carry them for half a mile or so, occasionally smashing them against trees and rocks. This was at first thought to be the work of hyenas, but hyenas could hardly drag tusks weighing 80 or 100 lb. for such a distance, let alone smash them against rocks.

Ear-movements and scent play a big part in the communal life of African elephants. Our knowledge of this is due in no small degree to W. Kühme's years-long study of the two bulls and one cow at the Opel open-air zoo near Frankfurt. These animals have a very spacious paddock and are not tethered at night. They have also bred in captivity. The ears are flapped back and forth with varying frequency depending on the air-temperature. Even a rise in temperature of as little as half a degree causes the ears to move with increasing rapidity. Aggressive intent is expressed by spreading the ears and moving them more rapidly than the air-temperature would normally warrant. When in the vicinity of higher-ranking animals, subordinates refrain from moving their ears at quite the speed warranted by the outside temperature. The ears are never flattened except in pure panic.

Elephants dig by going down on the ankle-joints of their forelegs and loosening the soil with their tusks. They enjoy a good roll and sprinkle themselves generously with dust which they sweep together with their forefeet and trunk and spray over their bodies with the latter. They also shower themselves with branches, leaves, grass, earth, and dung. They sometimes spread their ears and give their raised heads a short sharp shake, often with such violence that one of their forelegs echoes the movement. They tend to nod their heads before attacking or charging, and can turn on the spot for minutes on end.

Being primarily creatures of scent, elephants use their trunks to explore their own and other elephants' heads and, more especially, mouths, their temporal glands and ears. A bull will often apply the tip of his trunk to a cow's shoulder and intermittently convey it to his mouth for some seconds.

Mock battles among bulls

An elephant which is entirely at peace with the world lays its ears flat and allows its trunk to hang down. When in an excited and aggressive mood it puts its ears forward and raises its head slightly. The tip of its trunk rises a little, followed later, as excitement mounts, by the whole of the trunk. Finally, the trunk rears above the head, which is raised still farther. During a genuine attack launched in earnest, this posture changes once more. By contrast, an elephant which feels inferior folds its trunk back, inserting it in its mouth or grasping its own temporal gland or the edge of its ear.

Mock battles are common among bulls, especially young ones. Taking up their positions five or ten yards apart, the two contestants raise their heads, lift their trunks high, spread their ears, and charge until the base of one animal's trunk collides with that of the other. Shortly before impact the trunks themselves are intertwined or wound round the opponent's head. The animals then try to barge one another backwards, legs straining. If neither succeeds, the engagement is broken off. Both elephants retire and the shoving recommences a few minutes later. Attacks on human beings by half-tame elephants at Gangala na Bodio took the form of turning to face them, standing there for a moment with head raised, spreading the ears, and then charging at speed with the trunk hanging. The closer an elephant gets, the more its trunk swings or is curled towards the body. The tip of the trunk touches the left or right side of the head immediately prior to reaching an enemy and brushes the duct of one of the temporal glands just before a blow is launched. Trunk-strokes aimed at a human being are directed upwards or sideways.

Serious fights, which occur very rarely, are prefaced by the same routine but prosecuted with great violence and determination. In Queen Elizabeth Park, two bulls of very disparate size fought one another round a termites' nest with a small tree growing out of it. They boxed and butted their way all round and over the mound. Eventually the younger elephant tore out the small tree and brandished it threateningly with its trunk, but did not really use it as a weapon. There followed a series of collisions, head

When they fight in earnest

to head, with clashing tusks and trunks entwined. Although the larger bull was the first to sustain serious abrasions on its forehead, the younger lost its nerve and turned tail. It ran off, pursued by the other. Having followed the fugitive for some way, the big bull abandoned the chase and launched an attack on the watching game warden's field-car instead.

At Tsavo National Park, five bull elephants came down to the water to drink. Three of them dispersed afterwards, but the remaining two, which were heavily tusked, seemed to be in a fighting mood. One of them suddenly charged the other and drove it into the bush for about 80 yards. The evicted animal then turned and confronted its attacker. The elephants met head-on, and the aggressor, whose tusks were straighter, proved the more deadly of the two. Its right tusk pierced the roof of its opponent's mouth and its left tusk bored into the unfortunate animal's throat with such force that it was lifted off the ground. The attacker then withdrew its tusks, whereupon the wounded animal sank to its knees. It eventually fell dead after receiving another thrust through the shoulder which must have found its heart.

The victorious bull went back to the water and resumed drinking, not unhurt to judge by the numerous chest-wounds that were bleeding so profusely. It walked along the bank and, after taking another drink, returned to the field of battle. Seeing its late opponent still lying there, it attacked the carcass in a renewed access of fury and drove its tusks straight through the dead animal's skull, piercing the brain. It then rolled the huge carcase over until it faced in the opposite direction. Finally, it stationed itself beneath a tree about a hundred yards away.

One of the game wardens who had witnessed the duel went over to inspect the loser's carcase at close quarters but had to run for his life when the angry victor caught sight of him and charged. Shortly after this epic battle another herd of elephants approached the water but were promptly driven away and had to make a long detour before they could drink. The proud victor stood guard over the corpse for nearly six hours, then retired into the bush. The dead

The graveyard legend

elephant's tusks weighed 115 lb. and 109 lb., so it must have been a very large and mature bull. Its attacker was much the same size and had tusks of comparable weight.

Having one day shot an elephant in the Congolese province of Kiwu, Colonel R. Hoier found that its head contained the tip of an opponent's tusk, which had broken off against the skull. Tusks are often smashed in the course of such fights. The carcase of an elephant in the Sudan was left with a large section of its opponent's tusk impaled in it. It had broken into four pieces with a combined weight of 55 lb. This gives some idea of the force of such collisions.

Carrion-eaters converge from all directions when an elephant dies, especially vultures, which assemble in vast hordes. The flesh vanishes fast, but the bones are also pulled apart. The huge skull disintegrates rapidly, being largely composed of chambers enclosed by thin partitions of bone. It is untrue that one never comes across dead elephants, as some people allege. I myself have often seen the carcases of elephants that have died for reasons other than shooting. In recent years, 325 elephants were found in Murchison Falls National Park, all of which had died of natural causes and could be submitted to scientific examination. Nevertheless, the myth of the African elephant graveyards dies hard. According to this, elephants which sense that their time is at hand retire to marshes or remote areas where large accumulations of elephants' bones are later found. What prompted such myths may have been the mass-shootings of earlier times, also the practice of rounding up whole herds of elephants in tall grass and bush during the dry season and cornering them by deliberately kindling fires in which the animals die an agonizing death. Mass-poisoning by ivory-hunters may have been another contributory factor, also asphyxiation by pockets of gas in low-lying valleys. First identified by Dr Verschuren, these pockets are capable of killing off numerous large animals in quick succession. Thus, although elephants' carcases are normally found at scattered points throughout the countryside, the 'graveyard' legend may have been engendered by the accumulations of bones which are occasionally discovered at

Why elephants need no hair

individual places. The Turkana, for instance, used to set traps round isolated hillocks in the savannah, and killed numerous elephants in this way over the years.

Elephants are giants in the true sense of the word. They are the heaviest and, after the giraffe, the tallest land animals in the world. When dismembered and weighed in sections, male bush elephants have tipped the scales at $6\frac{1}{2}$ tons. Dr Laws, who examined 360 elephants from the Murchison Falls and Queen Elizabeth parks in Uganda, found that the average shoulder-height of bulls was 10 feet 4 inches at Murchison Falls and 9 feet 9 inches at Queen Elizabeth, the equivalent figure for cows being 8 feet 11 inches. The average weight of cows was 6,098 lb. and the record weight for a bull 13,230 lb. (nearly 6 tons.)

Circus and zoo elephants, most of which are cows, normally weigh 3-4 tons. Even a 5-ton elephant qualifies for the heavyweight class. In the case of the huge bull referred to above, the hide alone weighed 1 ton – 13 cwt. when dried – and covered an area of 377 square feet. The lungs weighed 302 lb., the liver $231\frac{1}{2}$ lb., the kidneys $39\frac{1}{2}$ lb., the ears 176 lb., the trunk $264\frac{1}{2}$ lb., the skeleton 3,527 lb., the heart 44 lb., the muscles 5,592 lb., and the fat 220 lb. The elephant carries far fewer square inches of hide per pound of body-weight than smaller animals. For that reason it loses comparatively little warmth via the surface, can do without hair, and tolerates our northern climate quite well. The ears, which in an African elephant constitute one-sixth of the surface of the body, are primarily used for cooling purposes. A largish elephant produces as much warmth as thirty human beings and its heart beats about thirty times per minute.

The tusks are not canines but converted incisors. There are African elephants which have no tusks at all, one tusk only, or a malformed cluster of between five and seven of them. Many tuskless elephants are found about 40 miles east of Chirundu in the Zambezi Valley in Southern Rhodesia – roughly 10 per cent, in fact. Elephants' tusks have no coating of enamel and continue to grow throughout their lifetime, but no more than 2 inches a year. The elephant has only two molars, one on each side of the jaw, but these are the size of a loaf of bread and may be renewed up to six times when their predecessors have worn out.

Tusks make their appearance between the ages of twelve months and three years. According to research carried out by Dr Laws in the Uganda national parks, the combined average weight of tusks in females aged about sixty is 39 lb. and in bulls 240 lb. The remains of molars that have become dislodged are either spat out or swallowed and excreted.

Undoubtedly the most remarkable feature of the elephant, anatomically speaking, is its trunk. The animal can use this elongated nose not only to break down trees, dig holes and kill other animals but also to pick up objects as small as a coin. An elephant's trunk holds $3\frac{1}{2}$-$4\frac{1}{2}$ gallons of water which can be squirted either into the animal's mouth or over its own body. Elephants can also bend their trunks in any direction and expand and contract them by means of the 40,000 bands of axial and annular muscles which they contain. The trunk is so tough and resistent, despite its boneless structure, that it is hard to sever with a knife. An elephant can nevertheless survive without one.

A one-year-old trunkless elephant calf was observed over a period of four months in the South Luangwa Game Reserve. It remained well-nourished even though it had to kneel down in order to pick up diospyros-fruit. In Addo Park an elephant with no tip to its trunk used to suck up oranges despite this disability and blow them into its mouth. An elephant shot in the Nzega district of Tanzania had a vestigial second trunk leading off the main one. This was 18 inches long and had only one finger at the tip.

The elephant's mobile trunk enables it to take scent high in the air like a giraffe or close to the ground like a wild dog. It is clear that elephants are guided principally by their sense of smell. This is borne out by the experience of a woman who tried to photograph an elephant in a dried-up Zambian river-bed while she herself stood at the foot of a tree on the bank. Having taken her first shot, she was astonished to see the elephant heading straight for her. She

Elephant brain-power

quickly slung the camera round her neck, dropped an apple which she happened to be holding, and prepared to beat a hasty retreat. The elephant moved gradually closer and halted on the other side of the tree. Very slowly, it put its trunk round the tree, picked up the apple and popped it into its mouth, then turned and walked off in the opposite direction. It was the scent of the apple which had led the animal to within a yard of the woman. In Kruger National Park, a badly injured elephant calf passed close to a game warden who was waiting beside a water-hole, then vanished into the bush. Two hours later a group of eight full-grown elephants turned up and started to drink. One large she-elephant began to sniff the air and examine the ground with her trunk. She came to the spot beneath the tree where the calf had been lying, explored the surrounding area, and then set off on the young animal's trail. She moved very slowly with her trunk almost touching the ground. Her sense of smell did not deceive her. An hour later she returned to the herd with the calf at her heels.

Pliny wrote 2,000 years ago: 'The elephant is endowed by nature with that which is seldom found among men, to wit, honesty, shrewdness, a sense of justice, and obedience in the faith. As soon as the new moon returns, the elephants repair to the rivers. There they solemnly cleanse themselves and bathe, and, having saluted the planets after this fashion, return to the forests. When they are sick and lie down, they throw grass heavenwards as though they mean to make sacrifice.'

Similarly anthropomorphic and mystical views on the nature of the elephant are still to be found in the novels and adventure-stories of today.

An elephant's brain weighs $8\frac{1}{2}$-11 lb., or three times as much as that of a man, but little is known of the African elephant's intellectual capacity. In so far as I and other people have conducted experiments in this field, our research has all been concerned with the Asiatic variety. Working-elephants respond to more than thirty different commands and can deal quite purposefully and intelligently with tree-trunks, ropes, and so on.

Forty years is very old

It needed a lot of persuasion before I got a zoo elephant to stand on my foot. It was not squashed or injured but felt as if someone had dumped a 2 cwt. sack of grain on it. It would, of course, be inadvisable to insert one's foot beneath the leading edge of an elephant's foot, where the hoofs are situated. The rear end of the elephant's sole is covered with a gelatinous mass which accounts for its quiet and soft-footed gait. Even when a herd of elephants stampedes, there is no clatter or thunder of hoofs like that made by antelopes and zebras.

The elephant's foot spreads and expands under pressure but contracts considerably when raised from the ground – in fact the surface of the sole can diminish by about 25 per cent. This enables the animal to withdraw its legs from deep mud with ease. In 1958, a badly injured elephant in Kruger National Park was found to be standing on four raw stumps. Its charred soles lay not far away, having been burnt off when the animal got caught in a bush-fire on the other side of the Portuguese frontier.

Exaggerated statements used to be made about the elephant's life-span. Research carried out on 325 elephants that had died of natural causes in Uganda's two national parks indicated an average life-expectancy of less than fifteen years. The animals are probably shorter-lived than human beings. In the case of zoos, where individual specimens undoubtedly live longer than they would in the wild, an inquiry conducted by Dr Alfred Seitz failed to elicit a single instance in which an African elephant had exceeded the age of forty. (Most Indian elephants die before the age of fifty, and none has ever been known to reach sixty.)

Elephants need fairly large quantities of fodder because their utilization of food is poor. At Frankfurt Zoo we give our elephants 13 lb. of mashed oats, $6\frac{1}{2}$ lb. of bran and about 165 lb. of green fodder daily. Work-horses weighing about one-fifth as much consume up to 90 lb. of green fodder in addition to straw and 5–6 lb. of potato-flakes mixed with chaff.

Every beast at the Gangala na Bodio elephant centre is given a daily ration of 7–8 cwt. of foliage and greenstuff, but much of this is

What an elephant consumes

trampled and strewn around. When the residue was weighed, it turned out that each animal really consumed only about 3 cwt. of fodder and some 33 gallons of water in the course of 24 hours, though much of the latter was squirted over the body.

An elephant excretes its large cakes of dung, which are 80 per cent water, about once every hour during the day or fourteen to eighteen times in 24 hours. It has to get up during the night to excrete dung and urine. Despite its substantial intake of fodder, the elephant's capacity for work is not exceptional. A pair of elephants will pull a two-ton cart with a four-ton load – a total of six tons, or no more than their own weight – though the roads round Gangala na Bodio, which is the source of these statistics, are not exactly smooth. Harnessed to such a cart they can cover only $12\frac{1}{2}$ miles a day at a speed of $2\frac{1}{2}$ miles an hour. They are not permitted to work more than five days a week and must be rested and given food for about ten minutes in every hour. An elephant will carry 3-4 cwt. on its back and plough 0.8 of an acre to a depth of about five inches in four hours. However, these figures may have been deliberately scaled down to discourage the farmers who hire the animals from overtaxing them. The commandant told me that his elephants could actually haul up to nine tons.

The bush elephant of East Africa clearly favours a habitat consisting of grassland which also has trees or rivers fringed with trees to whose shade it can retire during the heat of the day. Irvin Buss analysed the stomach-contents of 47 elephants shot between January and March 1959 in terrain of this type just outside the borders of Murchison Falls National Park in Uganda. It turned out that 91 per cent of their food consisted of grass, 8 per cent of trees and bushes, and 1 per cent of herbaceous plants. Only about one-tenth of the grass was green and immature whereas nine-tenths were brown. The animals are constantly on the move while grazing. They become thoroughly engrossed in the joys of eating, particularly when consuming fresh grass that has sprung up after a bush-fire. One can then approach to within 15 yards without their noticing because they either close their eyes entirely while grazing

Eating habits

or focus them on the ground immediately at their feet. From this range, all that can be seen of their eyes is the upper lids and beautiful long lashes. Grass is plucked up by the roots and earth knocked off against a leg or the trunk. According to Nicholson, saliva often drips from the animals' mouths on such occasions. When the grassy plains are parched during the dry season, elephants spend a greater proportion of their time browsing on trees.

Huge as they are, elephants can be extremely fastidious in their tastes. The finger-equipped tip of the trunk which gives them such a fine sense of smell also enables them to pick out individual herbs and leaves at will. On its first nocturnal raid, a bull which cautiously climbed the fence enclosing the game warden's vegetable-garden at Paraa Lodge in Murchison Falls National Park ate all the green maize, melons, cucumbers and beetroots and some of the aubergines. On its second visit it polished off the aubergines and treated itself to some peppers. Tomatoes, lettuce, carrots and passion-fruit were left untouched.

In Tanzania the elephants generally make for bamboo thickets during October and November, feeding there for days upon an almost exclusive diet of bamboo-shoots. In February, when the mugongo-fruit are ripe, large numbers of them move south and gather in the Litou-Kiperere district. During the dry season they like to roam in areas where there is plenty of magugu-grass, which is broad-leafed and grows to a height of 20 feet. Elephants in Kruger National Park, South Africa, fall hungrily upon the fruit of the marula-tree as soon as they are ripe. These ferment inside the elephants' stomachs and intoxicate them. Some of the animals become aggressive, others pleasantly fuddled. Although tipsy elephants get up to all kinds of mischief every year, they do not present a hazard to visitors because the marula-trees have all been carefully removed from places frequented by tourists. The marula is one of the most widespread trees in Kruger. The natives brew an extremely potent beer from its fruit, which is why they refrain from felling it for firewood. The fruit ripen first at the north end of the park, so the elephants work their way slowly southwards, feeding

With a hut on its shoulders

as they go. One African game warden was knocked off his bicycle by a drunken bull. Another inebriated elephant thrust its head into a native hut and ran off into the bush with the hut draped round its shoulders. Yet another bull dragged an equally tipsy African from his bicycle and carried him into the bush for about ten yards. According to the director of the national park, Dr N. van der Merwe, it then returned to the road and trampled the man's bicycle into the ground.

Elephant herds were once in the habit of moving around Africa in search of the best sources of food available at each season of the year. They are largely precluded from doing so today, which is why some of them adopt injurious feeding-habits.

Elephants are probably the only African animals to dig waterholes. They begin by loosening the earth with their tusks and then use their trunks to scoop out the holes, which are often steep-sided and several feet deep. In this way they make water available, usually in the sandy beds of dried-up rivers, for many other creatures including rhinos, antelopes, zebras, flocks of birds, and snakes. They thus help other wild life to survive the dry season, much as some kangaroo-rats do in North-West Australia. A. Gordon states that he has seen elephants in Wankie Park sealing off their waterholes with bungs composed of leaves, grass or elephant-dung and then strewing them with sand. If elephants entirely abandon an area for some reason in time of drought, the effect on other creatures can be catastrophic.

In January and August 1957, or during the short and long dry seasons, the biologist H. K. Buechner counted more than 4,000 elephants in the southern part of Murchison Falls National Park in Uganda. Once the rainy season began in mid-September their numbers rose to twice that figure because animals left the more wooded regions outside the park and made for the grasslands. Since then the numbers of elephants in this national park have risen to a remarkable extent because the steady growth of the human population in neighbouring areas has prompted more and more of them to invade the park and remain there throughout the year.

Northern and Southern elephants

In 1964, by which time the elephants had already inhibited the growth of trees over a wide area of the park, two hundred of them were shot as an experiment in the area south of the Victoria Nile and another two hundred on the north side. Careful study of the slaughtered beasts enabled scientists to discern clear differences between the elephants on the north and south banks. Northern females attain sexual maturity at an average age of fourteen (ten to fifteen), whereas the cows on the south bank do so at eighteen (thirteen to twenty-two).

The corresponding ages for bulls are thirteen (eight to fifteen) on the north bank and sixteen (thirteen to twenty) on the south side of the river. The average interval between births is four to five years in both elephant populations, but the average lactation period on the south bank is far shorter, almost certainly because the calf mortality rate is very high among these animals. Thus the period during which no new egg forms in the ovary is extended to eleven months on the south bank, compared with only four months north of the river. The elephant population on the north side is younger and on the increase, having an average age of nineteen. The southern elephants are twenty-two years old on average and decreasing in number. Animals under ten years of age constitute 40 per cent of the northern elephants but only 22 per cent of the southern. The cows on the north bank conceive between November and April, whereas conception among the southern cows has shifted to April-September. Nutrition is undoubtedly one reason for this. Between two and three times as many elephant calves were bred in 1963-4 as in 1964-5. The difference between the two elephant populations is attributable to the fact that far more damage has been done to trees and vegetation south of the Nile. Formerly a well-wooded or even forested area, it has been entirely transformed into savannah by the elephants. There are still a few areas on the north bank stocked with trees and bushes. The population density north of the Nile is 0.36 per square mile, south of the Nile almost 0.7.

In former times, when the whole of Africa still belonged to the elephant, a situation of this sort would have regulated itself. Either the grey herds would have left the denuded southern areas and moved elsewhere, or their numbers would have declined automatically as a result of under-nourishment. Today, however, there is a determination not to allow the national parks – the last remaining areas of genuine bush – to be converted into desert or semi-desert. This presents the authorities with a number of unpleasant decisions.

Forty-five thousand elephants used to be shot every year in the old days, computed on the basis of the 6,000-odd tons of ivory processed annually throughout the world. The majority of this came from Africa, but nobody took the trouble in all these long years to inquire more closely into these hecatombs of slaughtered giants or discover something about the elephant's way of life. We may take it for granted that no genuinely free-ranging elephants will exist outside the few national parks of Africa by the end of this century. It is gratifying, therefore, that when overcrowding and deforestation in some of the national parks necessitates the killing of a certain number of animals despite inner reluctance on the part of game wardens, their death generally contributes to the furtherance of research.

Irvin Buss, for example, has been able to show that the 'musth' or discharge from the temporal glands in the side of the head does not, at least in African elephants, have any connection with rutting or sexual maturity. The opposite assumption prevailed for centuries. Elephants killed between July and October showed no traces of 'musth' but almost all those which were shot after 20th October and before the end of March betrayed definite signs of this discharge from the gland mid-way between eye and ear. It could not be detected in a few very young calves, but most sexually immature bulls and cows showed every sign of 'musth' during this period, as well as calves and older cows which had just given birth. The activity of this gland may have something to do with the dry season and high temperatures. It is clearly unconnected with reproduction.

Mating behaviour

Buss also established on this occasion that elephants with exceptionally long and heavy tusks, both cows and bulls, were getting on in years and already chewing on their fifth or sixth molar.

The cow's invitation to mate consists in pressing her hind quarters against the bull's head and half looking round. The bull lays his trunk lengthwise along her back. While she resists the pressure, he nudges her slowly forwards with his tusks and the base of his trunk. This procedure has been described by W. Kühme, who has made a close study of mating behaviour among African zoo elephants. The cow suddenly makes off at a fast pace, swinging her trunk and tail and nodding her head, with the bull following on the flank. She veers slightly towards him and he appears to head her off. The two animals place their heads together, often raising their trunks in the shape of the letter S. When they press the bases of their trunks together, as in a mock battle, the bull naturally prevails. The elephants wind and unwind their trunks, each animal delicately exploring the other's head with the tip of the trunk. The nudging process begins again, the cow presenting her hind quarters to the bull with gradually increasing frequency. She also goes down on her knees and lifts her tail while he feels for her vagina with his trunk. When, after more chasing about, she comes to a halt, he abruptly mounts her. The penis is inserted for a few seconds only.

W. Poles watched some free-ranging elephants mating in the Luangwa Valley Game Reserve in Zambia. The bull took the cow's tail in his mouth, pressed the side of his head against her hind quarters, then moved along her flank, laid his trunk across her neck and gripped her by the opposite ear. The cow stood there until the bull released her. Once again, the mating act proper lasted for only about ten seconds. Afterwards the animals faced one another and raised their trunks in the shape of an S. Two other bulls of similar size, which had been standing not far away throughout the performance, came up and walked between the couple without being challenged by the first bull. The four then grazed peacefully together. Before long the cow started to offer herself to one of the

Year-round mating season

new-comers, but did so four times in succession without arousing any interest. The plain was teeming with elephants, and there were several fully mature bulls within a few hundred yards of the mating-place.

In South Africa, A. Lewin photographed two elephants mating under water. Only the cow's head and neck and the mounted bull's head and shoulders protruded from the surface. Bulls frequently touch cows and entwine trunks in an affectionate way without showing any inclination to mate. As we have already mentioned, elephants mate throughout the year and do not have a definite breeding-season.

In the Nuanetsi district of Rhodesia, Allan Wright saw a young bull – obviously the youngest of the group, with tusks weighing about 20 lb. apiece – moving swiftly through the herd. Whenever he approached one of the cows she stopped grazing and stood still,

Page 305:
In order to conduct our experiments with inflated dummies, we pitched camp beside the Munge, a small stream on the floor of the Ngorongoro Crater. The animals at Ngorongoro are particularly well inured to the presence of cars and people.

A maned lion cautiously approaching a dummy lion in the Ngorongoro Crater. Its preliminary behaviour is identical with that normally adopted towards a male non-member of the pride.

Page 306, above:
Very gingerly, the lioness investigates a fallen dummy. When the animals grew bolder and seized the outsider by its tail and ears, even putting out their claws to touch it, the air escaped. It did not smell particularly pleasant because I had inflated the plastic skin with exhaust-gases from my car. This finally convinced the lions that the peculiar creature was not one of their own kind.

Page 306, below:
Our dummy elephant's polythene hide was rather too pale in colour, so we plastered the thin-skinned pachyderm with mud from trunk to tail-tip.

Page 307, above:
This elderly bull elephant on the shores of Lake Manyara spent time seriously debating whether to attack the outsider.

Page 307, below:
Our dummy aroused the curiosity of a wild elephant in Manyara National Park, Tanzania. It inquisitively edged closer and closer, raising its trunk to investigate the stranger. The fact that cars were standing around did not bother it.

only to hurry off if he came too close. The young bull seemed to be trying each cow in turn. The other bulls were quite unconcerned, though they too stood motionless when the youngster came too close. The latter's penis was unsheathed and erect.

The seven cows were eventually herded together and ran off. It was clear that they had no real intention of running away, because they moved in a tight circle, trumpeting and squealing, with their calves running ahead of them. The bull brought up the rear, attempting to mount one of them. Bushes and small trees were trampled underfoot. The four other bulls still seemed quite unconcerned. Eventually the young bull succeeded in mounting a cow which continued on its way for ten yards and then swerved sideways without shaking him off. He dismounted after about two minutes and the cow rejoined the others.

It seems that bulls are only disposed to mate at certain times which vary according to circumstances. Clearly, though, a combination of circumstances must be present before they do. One of the chief reasons why so few African elephants have been bred in zoos is that most zoos keep Indian elephants with, at best, one or two African cows as a side-line. Indian and African elephants have never been successfully cross-bred. However, the elephant-taming station in the Congo has always kept at least a dozen cows of breeding age and one or two bulls – sometimes considerably more – since 1925. The animals live in their natural environment. Although tethered at night like zoo animals, they graze in the open for hours or whole days under the supervision of their carnac, only a few of them being hobbled. Despite this, no more than four births occurred in thirty years – all, strange to relate, in 1930 apart from a single late-comer towards the end of the 1950's. The gestation period of the African elephant is just on twenty-two months. At Basle Zoo both bulls covered the cow between March 23rd and April 4th 1964 and the latter gave birth on January 12th, 1966.

Page 308:
The famous lions which love to rest high in the trees in Manyara National Park, Tanzania, gaze down with very limited interest at the dummy elephant beneath them.

Setting a calf on its feet

During the birth itself the mother-to-be is often surrounded – protected, in a manner of speaking – by other cows. In December 1956 Frank Poppleton was fortunate enough to observe the aftermath of a birth at Queen Elizabeth Park, where he was then game warden. He saw part of a largish herd of elephants crowded together, all facing outwards, trumpeting, flapping their ears, and behaving in an extremely restive manner. Poppleton was able to watch the proceedings through binoculars at a range of twenty-five yards. In the middle of the close-packed group was a new-born calf. The mother and another cow were engaged in removing the membranes. The mother's belly was enormously distended and hung down almost to the ground. The vaginal entrance, which in elephants is situated low and between the legs, not high and to the rear, was dilated and bleeding. The group consisted of six large cows and five small calves, together with a young bull which looked on from a distance of about fifteen yards.

Some of the animals were using their feet and trunks in an attempt to set the new-born calf on its legs. Others took the membrane and tossed it into the air so that it spread out like a sheet. The cows drove the vultures away and kept all elephants save the one bull at a distance. The calf was wet and covered with mucus. It had a perceptible coat of hair, particularly on the head. The general trumpeting and screaming persisted for about ten minutes.

After half an hour four of the cows and their calves moved off, leaving the new-born calf alone with its mother, another adult cow, and a young bull aged about seven. The rest of the herd also moved down to the river and disappeared. The mother, the other cow and the seven-year-old bull pursued their attempts to get the baby on its legs. After another fifteen minutes the second cow also departed. This left only the calf, its mother, and the young bull, which was obviously her son by a previous mating. The bull inserted its trunk beneath the little one's belly from the side and also between its legs, to help it up, but it was still very weak and repeatedly fell over. The mother continued to throw the membrane

The birth of an elephant

about and finally tossed it on to her back, where part of it clung. She made two attempts to devour the membranes but removed them from her mouth.

The baby elephant took its first wobbly steps two hours later but tumbled head over heels and rolled on to its back. Its mother and half-brother would not allow it to lie there, however, and set it on its feet time and time again. The cow, which continued to bleed persistently, expelled the placenta two hours after giving birth. She devoured some of the outer skin but left the greater part and spent some time playing with it. For a while, it dangled from one of her tusks. The young bull lost interest after two hours and vanished in the direction of the herd. When Poppleton approached in his field-car the mother immediately charged him.

An actual birth was witnessed in Zambia by a game warden who had spotted a she-elephant leaning against a tree. The youngster's head emerged from her vagina, and then, after steady pressure, the whole calf tumbled to the ground. The cow at once turned and remained standing over her offspring for twenty or thirty minutes, sniffing it with her trunk. The baby elephant then rose to its feet and started to look for the udder. It stood at right angles to its mother, with its forelegs planted in front of and to the rear of one of her forefeet, and began to drink. No other elephants were in the neighbourhood throughout this time. By contrast, R. Hoier was once stuck in his car for four hours in the Congo because a cow had just given birth beside the road near a small bridge. No amount of shouting and tooting could induce her and her companions to budge.

At Etosha National Park in South-West Africa, a young cow was observed giving birth while the rest of the herd grazed in a semi-circle about 400 yards from her. The nearest elephant, which appeared to be an elderly cow, was grazing about 200 yards downwind of her. The young cow stood there almost cowering with her hind legs bent, and trumpeted piercingly as though in mortal fear. Game Warden Baard got out of his car and went to inspect her at closer range. When he got to within about 20 yards of her he saw

Births in the Basle Zoo

her muscles suddenly contract, beginning at the ribs and rippling across the belly. Then she started to trumpet again.

The remainder of the herd took no discernible notice, though the older cow did raise her trunk. The cow under observation was smitten with painful spasms at intervals of about five minutes and trumpeted on each occasion. After thirty minutes, by which time the cow seemed very exhausted, the calf's head appeared. It was another fifteen minutes before the young animal's forelegs and shoulders emerged from the vaginal entrance. Suddenly the cow sank to the ground and lay on her right side, trumpeting at shorter intervals now. At last she emitted a long groan and lay very still, breathing slowly.

The new-born elephant kicked and struggled on the ground in an attempt to free itself from the egg-membrane, but the mother was either unconscious or too exhausted to bother about her offspring. The calf, which had meanwhile almost extricated itself from the membranes, was pink in colour except for the soles of its feet, which were yellowish-brown. After about ten minutes the cow raised her trunk and pointed it in the calf's direction. One hour and ten minutes had elapsed between the onlooker's arrival and the birth itself.

Much the same happened when the young cow Idunda gave birth to her first calf at Basle Zoo in January 1966. One morning the keeper discovered a large, tough, brownish plug of mucus, and at 9.15 a.m. next day the first pains made themselves felt. The cow felt her swollen udder with her trunk, also, from time to time, her belly and somewhat enlarged vagina. She lay flat on her side and rolled back and forth. The pains were unmistakable and recurred every four to six minutes, accompanied by the excretion of progressively smaller quantities of dung and urine. The first amniotic fluid began to flow at 9.36 a.m., and thereafter the area beneath the anus became peculiarly swollen.

The cow became very restless. She got up and paced to and fro, flapped her ears, braced her tusks against the ground and almost stood on her head. At the same time, her udder perceptibly in-

Mothers and their calves

creased in size and the dugs swelled. She repeatedly lay down. At 9.55 a.m. the bag of waters became visible and shortly afterwards burst. The mother-to-be then flexed her hind legs slightly and the calf emerged, hind legs first. The process was so rapid that it could not be followed in full detail with the eye, only the camera. After a few minutes the mother felt her baby all over and used her trunk to lift it slightly by the tail. Another fifteen minutes and the little elephant could stand, though only with the keeper's assistance. It weighed 249 lb. and was 3 feet 1 inch tall. The youngster did not find its mother's right dug until nearly eight o'clock next morning, when it drank copiously. African elephants had previously been born in Germany during the last war in Munich and in August 1965 at the Opel open-air zoo in the Taunus.

The Congolese elephant centre gives average figures of 220 lb. weight and 31-$33\frac{1}{2}$ inches height at birth. At one year the young elephant is a good 39 inches tall, at two years $43\frac{1}{2}$-47 inches and weaned. It grows 4 inches annually until the fifth year and somewhat less thereafter. The tusks, which break through between one and three years after birth, are easily visible at the age of five.

Young elephants are extremely clumsy in the early stages and rely on their mothers' assistance. When a group of cows with calves were scaling a steep slope beside the Nile near Fajao, one very small calf simply could not manage the gradient. Its mother knelt down at the top of the bank, grasped the little creature with her trunk, and very gently pulled it up. On the Kolozi river in Zambia, the game warden Kachari encountered two she-elephants and their calves. The animals ran away up a hill, the leading cow constantly urging her calf to climb faster by nudging it with her trunk. When they reached a place where there were numerous fallen trees, the same cow lifted them so that her calf could proceed without difficulty. Whenever it reached an impasse, the mother raised the tree-trunk to allow the calf to pass beneath.

While accompanying the Minister of Natural Resources to Queen Elizabeth Park in 1959, R. M. Bere came across a she-elephant carrying a dead calf. From the stench, it must have died three or

Cult of the elephant

four days earlier. The mother progressed slowly because she never deposited the corpse on the ground in order to graze or drink, but the other elephants in the group regularly waited for her. Although she used her tusks to pick the carcase up, she carried it between her lower jaw and shoulder 'like a violinist holding his instrument'.

All in all, we still know regrettably little about the African elephant's needs and way of life. What makes this even more regrettable is that we shall have in the next few years to find ways and means of helping at least a proportion of these noble and majestic beasts to survive the mounting pressures of civilization and human population density. They are the true kings of the animal world because they fear no natural enemy, not even the lion.

Such is the view of the Dan, a tribe found in parts of Liberia and the Ivory Coast. I have visited these people on two occasions. As Hans Himmelheber's research has shown, they place less emphasis on man than on the spirit that dwells within him. The same spirit can simultaneously inhabit a man and an animal. Anyone who senses his ability to dwell in spirit simultaneously in a man and a particular animal must seek admission to the appropriate animal society, and often has to pay the master of that society a considerable sum. The man whose spirit can also be that of an animal develops the special characteristics of that species of animal in his human body.

The supreme animal society is that of the elephant, whose members are masters of men, notably tribal chiefs. These individuals become tall of stature, develop a massive paunch, and walk with ponderous tread. A chief's spirit does not enter an existing elephant. Instead, it creates for itself a new one which did not exist before. If this elephant is killed in the bush, the man in question is bound to die too. Dying men often explain that they must depart this life because they also happen to be the elephant or buffalo which was killed yesterday or the day before. When an important person dies for no ascertainable reason, his death is generally attributed to sorcery and the miscreant has to be identified by witchdoctors. Before this unpleasant form of investigation gets under

Taboo districts and national parks

way, the corpse is carefully undressed and examined to see if there is any abrasion or small ulcer which may point to a wound in the corresponding animal. The hunter who killed it is not punished because he was only doing his job and supplying the village with meat.

Admission to the elephant society is the most expensive of all. A well-to-do applicant suddenly incurs debts, cannot afford to buy anything, mysteriously loses five of his ten cows. Although he discloses nothing, his neighbours exchange whispered gossip to the effect that he is raising the entrance fee to a spirit society. If he can eventually announce, beaming with satisfaction, that he is now 'in the elephant', his stock goes up by leaps and bounds. Everyone falls silent when he joins in a discussion. He is also favoured for election to minor offices or even the chieftainship because everyone knows that the elephants are behind him.

For this reason, the Dan consider laws for the protection of wild animals pointless. All animals now living in the bush are secondary forms of human beings, after all, so they cannot become extinct as long as the human race survives...

On the other hand, the ancient African religions have often worked in favour of wild life. Most regions have their tabu areas, sacred groves and tracts of countryside where hunting is prohibited. Religious laws of this type are very conscientiously observed because it is known that the gods or spirits are implacable in their punishment of offenders, however discreet. Thus, if a district was hunted too intensively and many species of game disappeared, new stock gradually infiltrated the neighbourhood from the tabu areas. The more the old religions of Africa wane, the less people worry about sacred prohibitions. For the benefit of modern Africa, therefore, tabu areas must be replaced by national parks.

17. *The Nile, frontier between species*

*There is more to a woman than her sources of milk; the goat also has two.
If you fear, fear men. Animals pass by.*

AFRICAN PROVERBS

Tall stories told by big-game hunters sometimes take a century to explode. Mr Cornwallis Harris, who toured the North-West Transvaal in 1836, described how he encountered no less than twenty-two white rhinos during his trek from the Limpopo and had to shoot four of them 'in self-defence'. In recent decades we have taken a closer look at what is, after the Indian and African elephants, the third-largest land animal in the world. Dozens of observers with less itchy trigger-fingers than the hunters of old have taken the trouble to watch them day and night, and in recent years they have been captured, transported to other areas, and released. Today, nobody would accept the African explorer's plea of 'self-defence'.

We are lucky to be in any position at all to study the true nature of these mighty creatures, so nearly did the heroic hunters of the past come to annihilating them. Nichols and Eglinton wrote in 1892 that there was every reason to suppose the white rhinoceros extinct, and Brydon reiterated this statement in 1897.

For all their vast bulk, white rhinos are so markedly placid that they would have made a far more appropriate United Nations symbol than the quarrelsome dove. They also get on far better with other large animals like elephants and Cape buffalo than their smaller and more unpredictable cousins the black rhinos. At Nimule National Park in the Southern Sudan, white rhinos have been seen quietly sharing the shade of the same tree with elephants. Similarly, white rhinos very seldom subject human beings to the startling exploratory sallies favoured by their 'black' relatives, which make sudden charges, pull up at a distance of 5-8 yards, snort, shake their heads, and then walk away. In areas where they are not hunted, white rhinos generally allow people and cars to within 30 yards or so before making off. One night in Uganda, J. B. Heppes watched a white rhino pass quietly between his tent and camp fire, which were only 10 yards apart, on its way down to the water to drink. It returned by the same route without bothering about the fire or its proximity to man.

I know of only three cases, reported by the elephant-hunter C. H. Stigand, in which white rhinos have killed human beings. In the first, a woman picking cotton surprised a rhinoceros cow with her calf and was attacked and killed. In the second, a similar fate is said to have befallen a man who was shouting at a rhinoceros which had invaded his field. Finally, in Zululand, a group of yelling natives drove a rhinoceros from a protected area down a narrow path. When the panic-stricken beast met a woman coming the other way, it ripped open her body with its horn and trampled her to death. During a catching-expedition in Uganda, on the other hand, a man fell off the back of a truck which was being hotly pursued by a white rhinoceros. He was able to catch the truck as it jolted along and climb aboard without being injured by the furious animal. It is interesting to note that in Germany, prior to the mechanization of agriculture, four hundred people were killed each year by farm cattle and horses. Thus, shooting a white rhinoceros is probably just about as dangerous as shooting a domestic cow from inside the fence rather than outside.

Be that as it may, a white or square-lipped rhinoceros measures almost 6 feet at the shoulder and $13\frac{1}{2}$ feet in length, and its head alone is 4 feet long, or about 25 per cent longer than that of the black rhinoceros. It also weighs up to 4,400 lb. in comparison with the black rhino's 2,000-3,000 lb. It has three toes like the latter but its spoor is almost twice the size. The longer fore-horn averages

Mock battles horn to horn

31 inches in length, though the longest on record measured 63 inches. The horns weigh 14-20 lb. The animals can do approximately 18 m.p.h. at a trot and often attain 25 m.p.h. at a gallop. It is well known that white rhinos are no more white than black rhinos are black. The illogical designation 'white' may be a misinterpretation of the Boer word 'wijd', which corresponds to the English 'wide' or broad and is pronounced in almost the same way. Alternatively, it may derive from the native expression 'white heart', which implies peaceable. The white or square-lipped rhinoceros has no prehensile 'finger' or protruding upper lip. The function of its mouth is to crop grass, whereas hook-lipped rhinos do more browsing. White rhinos see about as poorly as black rhinos.

They are, however, far more sociable than their related species. I have often met herds of ten, fifteen and even twenty head in the Umfolozi Game Reserve in Natal. In contrast to their 'black' relatives, they invariably carry their heads close to the ground, cropping the grass until it looks mowed. Their piles of dung are also much larger than those of the black rhinoceros. My associate Dr D. Backhaus spent weeks at Garamba National Park in the Eastern Congo observing pairs of males and females. There, too, the animals fought mock battles horn against horn, but only in cool weather and only for a few minutes at a time. The contestants might be two bulls, two cows, or a bull and a cow. They once drove off two half-grown elephants and were not alarmed when starlings rose from their bodies at the approach of a car. The bulls sprayed the bushes with urine. Each animal deposited its dung on a permanent heap. A herd covered only 1,000 yards in seven to eight hours of grazing. The animals normally rested for two hours in the morning, from 10 a.m. until noon. White rhinos were not affected by the outbreak of rinderpest which raged during 1953-4.

Bulls fight ferociously, sometimes with fatal consequences. Wilhelm Schack, a former keeper at Frankfurt Zoo who later became a successful animal-watcher and photographer in Africa, once saw two bulls straining shoulder to shoulder during a serious fight, each clearly trying to prevent the other from using its horn to full effect. In the case of one furious encounter, 43 white rhinos drifted up to watch. A bull rhinoceros in pursuit of a cow usually lays his head on her back. Once he has mounted, the act of copulation may take up to an hour. The animals are reputed to have a gestation period of 540-550 days, but we have no detailed information about that or the birth itself because white rhinos have never bred in zoos. There are two sightings of twins on record. A calf can follow its mother twenty-four hours after birth and begins to graze when it is about a week old, though it continues to suck for at least a year. We know that baby rhinos have a full coat of hair by the age of four months thanks to a little white rhino which was found beside its dead mother and taken to Pretoria Zoo. One cow at the Umfolozi Game Reserve gave birth to a calf at the advanced age of thirty-six. Female rhinos are said to become sexually mature after three years and to produce young at three- to five-yearly intervals.

William Burchell was the first man to discover these huge but placid animals in the northern part of South Africa's Cape Province in 1812. They were extremely numerous in those days, to judge by aboriginal cave-drawings in South-West Africa, Rhodesia, Botswana, and Cape Province. These pictures are unmistakably of white rhinos. The white rhinoceros used also to be found in the south of what is now Angola, in part of South-West Africa, Botswana, Rhodesia, the Transvaal, Zululand, and in at least part of Portuguese East Africa. It is generally assumed not to have occurred south of the Orange River. Even in historical times, rhinos probably lived in the extreme south-west of Malawi (Barotseland), in the area between the Mashi and Zambezi rivers. It took us trigger-happy Europeans only a few decades to dispose of these harmless giants, yet it still came as a surprise when they suddenly vanished altogether like the quagga and blue-buck. People were glad to hear that a few dozen had survived in Zululand (Natal) in the lowlands between the Black and White Umfolozi rivers. In 1922 F. Vaughan Kirby announced that there were only twenty

Re-introduced in South Africa

left and appealed publicly for the survivors of the species to be given protection.

It is probable that Kirby deliberately understated their numbers, because in 1932 180 of them were counted in the Umfolozi area and another 30 in the neighbourhood. By 1948 their numbers had risen to 550 and they had also spread to the Hluhluwe reservation 15 miles away. The whole area, like many other regions of South Africa, was then cleared of tsetse-flies by spraying it with D.D.T. from the air. As a result, farmers and, in particular, local natives began to take an interest in the land they had earlier shunned, especially the territory between the two reservations. It was not long before renewed complaints were heard about the rhinos' numbers. By 1965, just on 1,000 of them were living in the Umfolozi (112 sq.m.) and Hluhluwe (61 sq. m.) districts, which is why some of them have in recent years been given away to zoos and other African reservations. For decades, the white rhinoceros remained unrepresented in any zoo in the world with the exception of Pretoria, which acquired first one and then a pair, but by 1963 the world's zoos already had 32, 13 of them in Europe. By April 1965 there were 25 white rhinos in the United States alone. Great experience has already been gained in transporting the huge beasts. By 1966, a total of 150 white rhinos had been reintroduced into other South African national parks and reservations, and another dozen were sent to Rhodesia. They have since been resettled in the Kruger, Wankie and Matapos National Parks, where they used to occur in earlier times, but also in entirely new areas such as the Kyle Dam reservation. Some have even travelled to Kenya's newly established (1966) Meru National Park in East Africa.

But who would ever have dreamed of finding the huge square-lipped rhinoceros north of the equator too, almost 2,000 miles from its southern habitat? People were reluctant to believe it at first when, in 1900, Major A. Gibbons discovered some white rhinos in Uganda on the left bank of the Upper Nile. My Upper Silesian compatriot Emin Pasha – alias Dr Eduard Schnitzer (1840-1892) – had spent five years there when cut off from all contact with the outside world by the Mahdi rebellion. Was it possible that such a keen student of wild life had overlooked so conspicuous an animal? Schnitzer, who was alone among the great explorers of his day in treating the natives with humanity rather than condescension, had probably concentrated too much on the birds and small game that were his real love in life. At all events, it soon emerged that the northern square-lipped rhinoceros grazed on the left bank of the Upper Nile between it and the tropical rain-forest, in Uganda, the Sudan, the adjoining Congo, and far into the French territory of Ubangi.

Forty-sixty head were counted in the Belgian Congo in 1925. In the same year, especially for them and for the only giraffe then known to be resident in the area, there was established the Garamba National Park, which adjoins the Sudan in the south. Their numbers increased steadily after that and may now stand at about 1,000 provided not too many have been shot by poachers in the turmoil of civil war. In French Equatorial Africa, by contrast, they were probably extinct as early as 1931. A few of them also lived in Nimule National Park on the Sudan-Uganda border. Northern white rhinos have never advanced eastwards across the Nile. Because of minor differences in bone-structure, they are regarded as a subspecies (Ceratotherium simum cottoni) distinct from the southern square-lipped rhinoceros (Ceratotherium simum simum).

It unfortunately proved impossible to afford effective protection to the rhinos in Uganda west of the Nile. They were persistently preyed upon by poachers, all the more so because of a steady rise in the human population. Only 60 are reputed to live there now. It was consequently decided in 1961, after much weighing of pros and cons, to capture some and transport them to Murchison Falls National Park, most of which lies south of the Victoria Nile. Opponents of the plan objected that white rhinos had never existed there – that the Nile had always been a virtually insurmountable barrier to the spread of black rhinos westwards and white rhinos eastwards. On the other hand, it was known that both species co-

Should black and white rhinos mix?

exist peacefully in the Hluhluwe reservation in South Africa. Although the mountainous northern part is inhabited exclusively by black rhinos and the flatter country by white, it was hoped that both types would tolerate one another in Murchison Falls Park too. This hope proved to be well founded.

In 1961 the first two white rhinos were captured with ropes and nooses and released in Murchison Falls Park. They were sighted again 39 days later, 10 miles from their point of release. Altogether, ten animals were transferred to the park over a period of some months, though two of them died a few days after release, probably of injuries received during their capture.

The only cow left a female calf which was artificially reared and could scarcely be prevailed upon to leave the game wardens' post at the airfield when it got older. This animal, which was christened Obongi, gave one visitor a terrible shock. He was busy photographing some antelopes in the distance when a rhino's head suddenly appeared between his legs and hoisted him into the air. The tourist was uninjured.

Even when dumped a considerable distance from the game wardens' huts, Obongi doggedly found her way back again and again by trailing the petrol tanker. She was later attacked and badly mauled by lions, whereupon she once more took refuge with her human friends. Eventually, when she was big enough to fend for herself, she remained in the bush. Obongi was occasionally sighted in the company of a bull rhinoceros. In January 1967, or almost two years later, she reappeared at the game warden's post in the Pakuba district and started to scrape the paintwork of visitors' cars as she used to in the old days. After six months she gave birth to a female calf. A year later she was wounded in the right shoulder by a poacher's spear but recovered. In 1969, game wardens found her nineteen-month-old calf wandering alone in the bush. Then they caught sight of vultures circling over the remains of Obongi herself. Poachers had crossed the Nile and killed her. Because the carcase had not yet been stripped of its horns, the game wardens laid an ambush. They failed to catch the culprits but

The white rhino preserved

managed to identify the number of their canoe, arrest them in their village, and convey them to gaol. Fortunately, the orphaned calf teamed up with another cow which had a calf of her own.

In 1964 another five white rhinos were captured. This time, game wardens hunted them in the bush by helicopter and guided field-cars to the spot. They were paralysed with injections of Sernyl shot into them by crossbow – some from the helicopter – and given oxygen. Most of them could stand again after four hours, though two cows in calf succumbed to the drug. It took eighteen hours to transport the animals across the Nile and into Murchison Falls Park, a journey of about 180 miles.

Today, we can breathe a sigh of relief at the thought that the white rhinoceros, almost extinct fifty years ago, has been saved by wild-life conservationists and is once more quite common in Africa.

18. The truth about the wild dog

Death is like the moon. What man ever saw its back?

AFRICAN PROVERB

Many stories about African animals have been stubbornly perpetuated, decade after decade, some of them for a century or more. African wild dogs – 'hyena-dogs' – are reputed to invade a district like a horde of ravening devils, dismember far more game than they ever devour, depopulate whole regions or, at best, drive all the grazing-animals away. No gazelle can escape them, it is said, because they hunt in relays. One dog pursues the quarry until its tongue is lolling with exhaustion and is then relieved by the next member of the pack. It is even alleged in a recent book that the animals killed a human being, a hunter who strolled out of his camp at the foot of Mount Meru and never came back. All that was found was five spent cartridges, five dead dogs, and some human remains. The rest of the pack had apparently devoured him. On the other hand, wild dogs have been systematically exterminated for decades, even in many national parks – e.g. South Africa's Kruger Park – by game wardens. The dogs are not particularly hard to kill because they have little fear of man and do not run away. The immediate result at Kruger was that the impalas multiplied inordinately and grazed the land bare. I myself have never heard of a single authenticated case in which African wild dogs have attacked human beings, far less killed them.

It is not unduly difficult to watch wild dogs hunting because, in contrast to many other wild animals, they have no very strong objection to a human audience. At Serengeti and in the Ngorongoro Crater in Tanzania, where game is abundant, they only hunt when the sun is just above the horizon, or between six-thirty and seven in the morning and between six and eight o'clock in the evening. Occasional exceptions do occur, of course, but the animals normally spend the daylight hours in their cool dens, originally the home of aardvark or wart-hogs. Alternatively, the whole pack divides into small groups which lie in the shade of separate trees. If one or other of the dogs feels an urge to hunt, it stands up and goes over to another group, encourages them to get up and starts to romp about. When all is in readiness, the dogs evacuate. One of them trots off in the direction of a group of Thomson's or Grant's gazelles, and the others follow suit. If the majority cannot be persuaded to move, the one or two would-be hunters retire to the shade and lie down again. Wild dogs never hunt singly.

Preying is done visually, never by scent. The animals seldom bother about wind direction and make little effort, unlike lions or leopards, to exploit cover. The pack walks or trots unobtrusively along, trying to get as close to its prey as possible but hardly ever following if the latter makes off before the range has closed to three hundred yards. Once the hunt is on, however, the leading dog or bitch will race after fleeing gazelles at a speed of approximately 35 m.p.h. A wild dog can maintain 30 m.p.h. for distances in excess of one mile.

The fugitive animal is seized from behind, usually by the leg or thigh but sometimes by the underside of the belly, and literally torn to pieces by the rest of the pack as they arrive. The kill itself takes a very short time, but the entire chase may last from three to five minutes and cover a distance of one or two miles. Prey have little chance of escape. Of 28 pack-hunts observed in the Ngorongoro Crater, 25 ended in the quarry's death.

As a general rule, predators select prey which will give them little trouble – in other words, animals which are either smaller or no bigger than themselves. Intense hunger may prompt them to tackle more powerful adversaries, but the risk to their own lives becomes correspondingly greater. More than two-thirds of the animals preyed on by wild dogs in the Ngorongoro Crater and

A horrifying spectacle

Serengeti Plain consist of little Thomson's gazelles and only 10 per cent of the larger Grant's gazelles, gnus and other sizeable grazing-animals. In the case of larger prey, the dogs simply rip open their bellies from behind so that the entrails fall out. The luckless creatures' hind legs fold up, leaving them almost powerless to fight off the dogs with their horns. It is a grisly and appalling spectacle. In the dogs' favour, it should be said that they are equipped solely for the chase and do not, unlike lions or leopards, have powerful paws barbed with claws or strong shoulders and neck-muscles which would enable them to break a victim's neck. Wild dogs are only sighted with prey which they have killed themselves, whereas lions are quite happy to accept the leavings of other predators. In South Africa, where game has become rare, the dogs are said to hunt for longer distances and pull down a larger proportion of big game. They are also somewhat larger and more heavily built than the East African variety.

During the dry season, only one water-hole remained in a wide area of the Luangwa Valley in Zambia. A pack of four male and three female wild dogs, together with seventeen young, occupied this solitary water-hole and picked off all the young antelopes and wart-hogs that came to drink there. They would not be dispersed by the game warden, R. G. Attwell, merely barked and retreated a few yards when he got out of his car or threw stones at them. It was not until he shot three of them that the rest of the pack came running up, nervously inspected the bleeding carcases, and finally withdrew. A minute or two later fifty impala and their young came down to drink a hundred yards away.

Bruce S. Wright kept a precise record of the daily kills made by wild dogs in the western part of the Serengeti Plain. They amounted to 0.33 lb. for every pound the pack itself weighed. Lions consumed 0.24-0.28 lb. per pound. The pack under observation killed 281 animals in a year, almost all of them Thomson's gazelles. Two-thirds of these were full-grown males. The probable reason is that males keep to themselves and lay claim to a particular territory. The larger a pack of wild dogs, the fewer the daily

Safe in the water

kills per dog that have to be made. A pack of 21 dogs killed 3.9 lb. per dog per day, whereas a small pack of only 6 dogs killed twice that weight. This is probably because smaller packs of wild dogs are often chased off their prey by hyenas. A singleton or pair would be unlikely to survive at all.

Once these flecked and long-eared carnivores are on the heels of an antelope, the latter has virtually no hope of outrunning them. There are, in fact, only two possible means of escape. The first is to plunge into a lake or river and – not that this is inevitable – run the risk of being taken by a crocodile. Wild dogs do not as a rule follow impala or waterbuck under such circumstances, and remain on the bank. At Mikumi National Park in Tanzania, J. Stephenson once saw a wild dog follow a male impala into the water and pursue it to the opposite bank. The impala turned and plunged in again, whereupon the dog followed, only to abandon the chase when the antelope swam close to three hippos. Watching from her balcony at Treetops Hotel in Kenya, Mrs Noel Tooly saw a full-grown waterbuck burst from the trees and plunge straight into the water, where it stood panting and trembling. Hot on its heels came twelve wild dogs. A few of them pursued it a little way into the water but soon gave up. The pack surrounded the pool for about twenty minutes and then withdrew. The waterbuck stayed put for a far longer time and then slowly emerged. A few of the wild dogs returned but the antelope dashed back into the water and menaced them with its horns. It eventually trotted off into the trees.

Strange and almost incredible as it may seem, another refuge to which many animals pursued by wild dogs turn in their desperation is man himself. I have recorded almost a dozen such cases in the past twenty years.

A game warden standing beside Luamfia Lagoon in the Chilongori reservation, Zambia, was surprised to see a young male kudu rush up to him and collapse with its tongue hanging out. The pack of wild dogs which had been pursuing it approached to within twenty yards and began to circle the game warden at a run. They then lay down and watched him for about fifteen minutes. The kudu slowly rose to its feet and pressed so close to the man that he could have touched it. When he threw some clods of earth at the nearest dog, the whole pack stood up and started to circle the pair once more. After a while, one of the dogs barked and ran off, followed at once by the whole pack. The kudu lingered beside the game warden for a minute or two longer and then disappeared into the tall grass round the lagoon.

Near Rugorogota in the Mbarara district of Uganda, a roan antelope pursued by six wild dogs ran up to the game warden Kartua Lorongsa, calling loudly. One of the dogs was hanging on to the antelope's tail, so the game warden shot it. At once, a second dog sank its teeth in the animal's tail. It was also shot, as were the third, fourth and fifth – the latter at point-blank range. The roan antelope, which did not budge from the game warden's side throughout this time, had only superficial wounds in its hind quarters. In the Parc National Albert a waterbuck sought refuge among groups of visitors but was pursued by wild dogs and killed notwithstanding. A similar fate befell a reedbuck which fled to C. Ionides but was torn to pieces by three wild dogs before his very eyes. Another full-grown roan antelope pursued by wild dogs, this time in Zambia, sought refuge on R. A. Critchley's farm and stayed there for five days until its wounds were healed and its self-confidence had been restored.

One year, three packs of wild dogs haunted the vicinity of Seronera in Serengeti. They numbered ten, eight and six head. On one occasion, two adult and five young dogs killed a Thomson's gazelle beside a car close to the camp at Seronera. On another, two large members of a pack consisting of fourteen adult and nine young dogs set off after forty gnus which were standing some eight hundred yards away. The bulk of the pack followed at a distance of two hundred yards. When a large hyena appeared, one of the dogs ran up, seized it by the hind leg and threw it. The hyena screamed but did not defend itself. Having approached to within four hundred yards of the gnus, the two dogs made a sudden dash and scattered

The rallying cry

them in all directions. When the dust had settled, small groups of gnus could be seen standing with their horns menacing the pack of dogs, which had now arrived in full force. The gnus lunged at them, but they played a waiting game. In the end an excited gnu calf broke ranks and was immediately killed. Its elders took little notice.

As soon as a pack splits up in the course of a hunt and individual dogs lose contact with the others, they put their heads very close to the ground and emit several bell-like howls. Then they raise their heads and listen intently. It seldom takes longer than five minutes for the whole pack to converge at a loping run.

The zoologist W. Kühme lived and slept behind the barred windows of his car for weeks on end at Serengeti so as to observe how the dogs behaved among themselves. Although quite remarkable, his findings have since been fully corroborated by two other students of the wild dog who also spent time at Serengeti and Ngorongoro, the biologists R. Estes and J. Goddard. In contrast to human beings, chickens, horses and many other gregarious creatures, the wild dogs of Africa subscribe to no proper form of authority or ranking order. They try to cajole rather than dominate or intimidate their fellows. A dog which wants something from another – a piece of meat, say, or its companionship – behaves in a thoroughly humble and obsequious fashion. When two groups from the same pack are reunited after an interval, or when the animals come to life after a rest, many salutations are exchanged. One dog licks another's face or nuzzles the corner of its mouth, grovelling with legs bent and head and muzzle raised. This is how young, sick and weakly dogs manage to obtain a share of the communal prey. They may even persuade others to chew meat for them and regurgitate it.

It is impossible to make a full study of the living habits of any species of animal in a zoo. The creatures must also, and more especially, be observed in the wild. On the other hand, many facts cannot become known unless fear of man wanes sufficiently to permit breeding in captivity. C. E. Cade, who built up and for many

Feeding the weak

years directed the zoo at the entrance to Nairobi National Park, was given an adult male wild dog which had been released from a trap. He kept it in a run next to some growing youngsters of the same species. After a while, Cade introduced one of these into the older animal's run. Their first encounter was fascinating to watch. The adult dog flew at the new-comer, which stood there quite still with its head erect and whimpered gently. Cade had never heard his wild dogs make such a sound before. This behaviour – this standing still and whimpering – neutralized the older dog's aggression. It braked sharply and refrained from doing the new-comer any harm. Cade witnessed the same procedure every time he introduced other half-grown whelps to the adult male. The only exception was a youngster who did not see the big dog coming, failed to go through the appropriate motions, and was bitten in the flank. It immediately froze, raised its head, and whimpered. Just as abruptly, the older dog calmed down. I know of only one instance, reported by A. Percival, in which a pack of wild dogs killed one of its number after the latter had been wounded. In general, sick and crippled animals which follow a hunt at their own pace and reach the prey minutes after the kill are readily permitted to eat with the pack.

Within a few days, Cade's young dogs joined the old male in forming a pack which consisted of three dogs and three bitches. In February, when one of the bitches first came on heat, she mated with the older dog. Neither of her two brothers showed any interest in her – in fact they avoided the pair – which made Cade wonder if mating couples normally withdrew from the pack when living in the wild. As soon as the bitch was unmistakably pregnant, he separated her from the others and put her behind a trellis-work fence through which she could see them. She became so restless and agitated that he put the big dog in with her. Then, because he feared that her mate would kill the new-born pups, Cade again separated them just before the bitch gave birth. She did so after a gestation period of 72 days. Two days later she started to pace up and down the fence which separated her from her mate with one of the pups in her mouth. It was clear that something must be done quickly because the little creature would not survive such treatment for long. Cade was chary of reuniting the bitch with the whole pack, so he experimented with the big dog only. To his astonishment, the bitch at once carried the pup to its father, deposited it on the ground, and began to lick it. The dog sniffed it but showed little further sign of interest. The scene which greeted Cade that evening was one of idyllic family life: the bitch and her litter lay in the nesting-box and the dog had taken up residence beside them.

African wild dogs were first bred in captivity at Breslau Zoo in 1930. In 1960, Bronx Zoo succeeded in rearing four out of a litter of six. There too, the dog was shut away at the time of birth but had to be reunited with the bitch because she became too agitated.

It is clear that all members of a wild dog pack can perform any function except that of suckling the young, though pups may be fed with regurgitated meat by all members, males included. One group which Kühme observed for weeks at Serengeti consisted of six dogs and two bitches. One of the bitches had eleven pups aged about three weeks and the other's litter was still in her den. After a successful hunt the older animals used to return to the den and regurgitate fresh meat into the mouths of the adult baby-sitters as well as the clamouring pups. Before setting out on another hunt the full-grown dogs would scamper up to each other with ears down and noses extended, lick each other's lips and even the bitches' dugs, and spray urine about like excited children. Extremes of delight made them throw themselves on their backs and kick their legs in the air. When greeting each other they emitted twittering, chattering sounds, particularly if excited, setting out on a hunt, pulling down prey or devouring meat. Their bark of alarm was short and deep-throated.

At Serengeti, Dr Hans Kruuk observed ten wild dogs of which four were very emaciated. The six well-nourished dogs raced into a dip in the ground and caught a Thomson's gazelle which they

Communal feeding of pups

killed and devoured. Then they returned to the four underfed dogs, which had taken no part in the hunt, and regurgitated the meat for their benefit.

One pack consisting of five dogs and a bitch spent almost the entire year at Ngorongoro in Tanzania. They regularly made excursions from the celebrated crater but never stayed away for long. It was possible to identify them because they had all been photographed from various angles and their ears were distinctively torn. At the end of February the only bitch in the pack gave birth to an equally ill-balanced litter of eight males and one female. The mother died on April 3rd. J. Goddard saw one of the dogs drag her carcase from the den, closely followed by the outraged pups, which were still trying to get milk from her. The pack now consisted of five males, but the adult dogs reared the pups by feeding them regularly with regurgitated meat. The young remained in the den, usually guarded by an adult, while the remainder of the pack went hunting. As soon as the full-grown dogs had killed they returned to the den. The pups thrust their muzzles into the corner of the adults' mouths and the adults duly regurgitated a quantity of meat. All the young were reared until they could trot after the pack, but many of them died thereafter. By the year's end, only four of the original nine pups were still alive, one of the casualties being the only female. Some of the little dogs used to trot along behind the pack for up to two miles, but little meat was left by the time they reached the kill. Now and again one of the adults would still deign to regurgitate some meat for them, but this happened less and less frequently as the pups grew older.

It is possible to make friends even with wild dogs because they generally take little notice of human beings and do not run away. One June, R. A. Critchley came across eight pups playing happily outside their den, an old aardvark burrow. Their mother showed no fear and readily accepted a big lump of meat which was thrown to her from the car. Their father was lying some 25 yards away. On later visits the bitch wagged her tail as soon as she saw the car coming. She sometimes had difficulty in finding scraps of meat

Wild dogs not so wild

when they were thrown – an indication that wild dogs do not possess a very highly developed sense of smell.

Margarete Trappe, a well-known shot and farmer of the pre-1914 era, ran two farms – Ngongongare and Momella – at the foot of Mount Meru. She is said to have encountered a large pack of wild dogs while out hunting with her own gun-dogs. 'Showing no signs of mutual timidity or hostility, the tame and wild dogs mingled and sniffed each other inquisitively without exchanging a single bite. Each party must gradually have lost interest in the other, because they quietly separated and trotted off.' Frau Trappe brought nine pups home, but five of them died next day. Two of the four that survived later lived with a British police officer at Arusha. They were completely tame and had free run of the town at first, but their habit of stealing poultry and taking the occasional bite out of people's legs eventually compelled him to chain them up. The pair left behind at Ngongongare readily allowed Frau Trappe and her children to walk them on the leash.

The celebrated British big-game hunter and explorer F. C. Selous (1851-1917) described an incident in which a wild dog was seized by a pack of hounds. The animal looked dead but was only play-acting. Selous was about to have it skinned when it jumped up and ran off.

Wild dogs sometimes take remarkably little notice of man. An instance of this occurred at Mikumi National Park, Tanzania, where a pack of 35 dogs rushed into camp at seven o'clock one morning in full view of the park's entire labour force, which had assembled for duty. The horde of dogs appeared like magic. They ran in all directions, scurrying between vehicles and buildings and passing within a dozen paces of the men, of whom they took no notice whatsoever. The air was filled with whimpers and howls, punctuated from time to time by a subdued monosyllabic bark. The turmoil lasted for two or three minutes, with each dog chasing or being chased by another. Then, to quote Game Warden Stephenson's account, order was suddenly restored because half the mob ran off in one direction and the rest in another. Although he did not witness the incident himself, one of his men informed him that the dogs had earlier been squabbling over an impala antelope. It was possible that the temporary confusion had resulted from two packs having converged on the same animal.

On another occasion five wild dogs took an obvious interest in a group of climbers on Kilimanjaro. 'When we set off again on the last laborious half-kilometre stretch, which took half an hour, they followed to the edge of the glacier, parallel with our own route, and showed a clearly perceptible interest in us. We found other tracks made by them which crossed our own. It was obvious that they, too, had visited the crater itself. This was at a time when the crater and the slopes of Kibo, the summit of Kilimanjaro, were carrying far more snow than usual. One could hardly conceive of a spot less suited to wild dogs. When we had reached the summit, altitude more than 6,300 metres, and were digging for the visitors' book, which was hard-frozen into the ice, the dogs were sitting on the glacier ice only a few hundred metres away at the same altitude as us, watching our every movement with extreme curiosity. They finally disappeared from view behind the rim of the glacier when we began to descend, and we did not see them again.' George Webb was nonetheless able to photograph the dogs in the snow at the very summit of Kilimanjaro. ('Tier' 1952, vol. 12, p. 13.)

Zebras and adult gnus show little fear of wild dogs and will even advance on them threateningly. Packs sometimes pester hippos or even elephants, but more in fun than anything else. At Katwe on the shores of Lake Edward, a pack of dogs silently assembled round a hippopotamus which was trying to reach the water. Some of them even sprang at the huge animal's chest and legs. Then, disturbed by the human onlooker, they formed a semi-circle round two elephants. The elephants trumpeted loudly in evident agitation and retreated.

The behaviour of wild dogs towards hyenas varies widely. At Vienna Zoo, one wild dog bit through the foot of a hyena in the next cage so that it hung by a shred of skin and the injured animal had to be destroyed. A spotted hyena pursued by eight wild dogs in Mikumi National Park took refuge under the game warden's field-car, which was parked outside his house. The wild dogs surrounded the vehicle but made off when they saw human beings in the vicinity. At Serengeti, four wild dogs which had seized a young hyena were driven off by a rescue party of eleven hyenas.

Small groups of two or three wild dogs could never survive in the Ngorongoro Crater, where approximately 420 spotted hyenas live, because they would regularly lose their prey to the larger animals. Even a pack of 21 wild dogs lost part of their kill to hyenas.

When the 'outriders' of a wild dog pack have killed an animal, they are frequently driven off their prey by waiting hyenas but recapture it as soon as the main body of the pack arrives. Hyenas sometimes loiter for hours among resting packs of wild dogs until a hunt begins. They may even pick their way between groups, devouring dung. Kühme actually saw one impatient hyena touch the face of a dog, 'whimpering in a friendly way'. Wild dogs fling themselves at hyenas which become too troublesome after a kill. A large hyena attacked by several dogs generally lies down, screaming and snarling, and snaps vainly over its shoulder. In rare cases, it simply lies down and abandons the struggle. Hyenas never take refuge in their dens, curiously enough, and even immature youngsters prefer to flee into dense undergrowth beside streams where the dogs will not follow them. Wild dogs have never been seen to kill or even seriously maim a hyena. On the other hand, hyenas frequently disperse wild dogs that have lingered over a kill provided the rest of the pack is out of easy reach. It is therefore crucial to the African wild dogs' survival to form packs of maximal size and, having done so, stick together.

Their habitat has shrunk in the last few thousand years. They were clearly and magnificently depicted by artists of the early Egyptian period (3000 B.C.), so they must still have been living in Northern Egypt at that time. Today, their northernmost area of incidence is the Sudan. Other animals which disappeared from Northern and Central Egypt at the same period include the elephant, buffalo, giraffe and rhinoceros.

19. Rhino tosses rhino

He who knows not the lion grasps it by the tail.

AFRICAN PROVERB

Yes, those were lions back there. I couldn't tell precisely how many – only their dark round ears projected above the grass. I removed my own inflated plastic lion from the roof of the Volkswagen bus and set it neatly on all fours in the grass. It looked comical, rather like an outsize children's toy.

Then I drove another 20 yards, switched off, and waited. The sun was slanting across the broad green expanse of the Ngorongoro Crater. 150 yards away, a few dozen gnus and zebras cocked a curious eye at the lions and me. Three crowned cranes glided silently out of the sky and landed between me and the concealed lions. Their yellow helmets of plumage looked like slender, luminous flowers. They made their way serenely through the grass, pecking here and there. The morning sun was just gaining strength. There wasn't another living soul within miles.

Time and patience were the order of the day – as ever, when animals are involved.

The lions took no notice of my bus. A vehicle of this type affords the best possible protection against wild animals. A fortnight before, an acquaintance of mine had been gored and killed by a she-elephant as he toured his game park on foot. I have never heard of anyone being killed in a car, even when, as sometimes happens, its body-work is pierced and dented by a charging rhino or elephant. Some other acquaintances, who had pitched camp almost at the spot where my own tent now stood, found a lion in the entrance of their tent one night and heard other lions clattering about among their saucepans. They set fire to the tent to drive the big cats away and jumped into their car, where they at once felt safe. The local lions were becoming more and more intrusive and playful, but I did not think they would readily attack a human being in earnest.

We students of animal behaviour have a hard time discovering what goes on inside animals' heads. Our position is much the same as that of the psychologist who would like to know what thoughts pass through the head of an unborn child in the mother's womb or an infant still unable to speak. Few human beings can recall, in retrospect, what they experienced prior to the age of three. Child psychologists have to find ways round this difficulty, just as we do. The psychologist R. A. Spitz, for example, held a man-sized doll wearing a face-mask over a baby. Any adult would have regarded the crude figure, with its rudimentary eyes, nose and mouth, as a grotesque and ghostly apparition, but the baby smiled up at it. In the same way, young and inexperienced ducklings or little fish will approach quite crude imitations of the parents they have never known.

Man is the only living creature which paints proper pictures. Anthropoid apes can wield brushes and paint, but they do not produce likenesses of themselves or other objects. That is why we tend to believe that man is also the only creature which can recognize itself and other things in effigy. In fact, primitive peoples such as many South American Indian tribes or the pygmies of the Congo were at first unable to recognize themselves or familiar objects in photographs.

Then again, one witnesses the most puzzling behaviour in animals or reads of it in letters. One of my correspondents stood a new oil-painting of a friend on his chest of drawers. His dog took one look at it, sprang up as though bitten by a tarantula, slunk beneath his chair, and barked at the portrait with fur bristling and teeth bared. Assi, a dog belonging to Dr Brigitte Scheven of Göttingen, looked up at a dog-breeders' magazine which its mistress

Dogs and TV

happened to be reading, scratched at the photograph of a boxer and then proceeded to sniff the head and hind quarters of every picture in turn. When a dog fell into a hole on television and started whimpering, Assi whimpered in sympathy and went searching for it behind the TV set. Some budgerigars spent all their time near two embroidered parrots on a curtain and tried to nibble the outlines. A male blackbird fought its own reflection in automobile hub-caps, pecking at it 40-50 times a minute, 16,000 times a day, for 24 days on end. Eibl-Eibesfeldt deceived fish among sea-coral with their own reflection. At Datschitz a stork swooped down and fiercely attacked a metal stork standing in a garden. It fought with such fury that it hurt itself on the metal beak and eventually fell to the ground exhausted.

These are rare cases, however. Animals do not in general take much notice of pictures, figures and reflections, but are we so different? How many times a week do we really look at the pictures hanging on our walls at home? Domestic animals which share our life among pictures, newspapers and TV sets do not make ideal subjects for such experiments...

Action at last! A big male lion rose from the grass and walked calmly over to my imitation lion, followed a few paces to the rear by another. Both animals had splendid manes and looked majestic in the extreme. They stared fixedly at the strange 'lion' as they approached, never taking their eyes off it for an instant. The other lions became curious and raised their heads. I could see them now, two lionesses and seven cubs, three of them quite small. The zebras in the vicinity, which had started to graze again, also raised their heads to watch like spectators in a stadium.

About thirty yards short of my balloon lion, the two real lions halted and stared unwaveringly into its painted eyes. They stood there motionless, their keen gaze fixed upon it. Few creatures have a more awesomely Jovian countenance than the wild lion, with its fringe of flowing mane. The two animals continued to gaze at the peculiar object. It seemed an eternity to me, but less than five minutes elapsed by my watch before they lay down, first one and then the other. Even then, they did not take their eyes off the stranger. The greater part of a lion's existence is devoted to lying and watching. These lions presumably intended to wait and see whether the unfamiliar creature did something, and, if so, what. But the balloon lion did not stir.

After another six minutes the yellow dignitaries got up and walked, very sedately, towards the dummy lion. Then they lay down again, their keen clear eyes still focused on it. I had seen maned lions from different prides greet each other in the same way – appraising stare, fearless gaze, gradual advance. It was a form of optical threat or intimidation. If a lion failed this test of nerves, it normally turned and walked away.

But my dummy lion did nothing of the sort. It stood there trading stare for stare. Did the two male lions really see it as a rival lion, or just a strange and unfamiliar object? Some francolins were courting less than thirty yards away. They took absolutely no notice of the lions, genuine or fake. Lions were of no concern to them because partridges are not lions' meat.

The pair now stood up, circled the dummy and approached it cautiously from the rear, sniffed its tail and flank but did not touch it. The lionesses and cubs were still some distance away. Then along came a sudden gust of wind and the flimsy plastic lion fell over. The two males retired twenty yards and lay down again, still gazing fixedly at the dummy. I drove up, parked the bus so that the watching lions could not see me get out and resurrect my decoy, placing a couple of spanners on its plastic feet-flaps so that the wind would not lift it again. At that moment the two lionesses and some of their cubs stood up and came trotting towards me. I jumped back into the bus, slammed the door, and drove a short distance so as to watch the next episode.

The lionesses were interested in the dummy lion, not in me. One of them bounded towards it. Simultaneously, the two lions jumped up and headed her off. They were so annoyed that they chased the offending lioness for 150 yards or so. Perhaps they had mistaken the dummy for a rival after all.

More dummy experiments

Meanwhile, the other lioness had arrived with the cubs. The stranger was submitted to a close inspection. There was more sniffing all round, and one of the lionesses gingerly took the tail in her mouth and pulled. The weakling of a dummy lion toppled over. The lionesses patted it cautiously with their paws, and one of them seized its ear in her teeth and dragged it a short distance through the grass. The two male lions remained aloof. One of the lionesses eventually tugged at the dummy, this time with her claws extended. I discovered later that she had made four small holes in the plastic skin. The escaping air could not have smelt very pleasant because we had inflated the dummy with exhaust fumes from the car. After twelve minutes the whole pride moved off without inflicting further damage on the artificial lion, which slowly shrivelled up as the air escaped.

That afternoon I chased some hyenas off the remains of a zebra which they had killed. Then I loaded part of their meal into the bus and drove round in quest of lions. In due course I sighted a lone lioness. Placing the haunch of zebra so that the wind carried its scent to her nostrils, I stationed my dummy lion precisely between it and her. With mingled longing and caution, she gradually approached. She did not venture to contest the unknown male's right to his kill or share it with him straight away. Instead, she wormed her way forwards on her belly, waited a while, crept still closer, waited again, then very gingerly took the edge of the meat between her teeth and pulled it sideways. Two other lions which I had failed to spot earlier came running up to share the meal.

So the lions had treated my dummy essentially as if it were

Page 341:
The cow rhinoceros, which had a calf to defend, attacked the dummy rhino in earnest. She impaled it and tossed it high into the air, then trotted off with her calf in tow. Soon, all that remained of the dummy rhino was a thin plastic skin.

Pages 342-3:
I found it a rather uneasy sensation, confronting an aggressive bull rhino with nothing but a balloonful of air between his horns and my ribs. It soon turned out that the rhino felt almost as uneasy as I did.

Alexander the Great was right

genuine. It may have resembled a lion in appearance, but it certainly carried no trace of lion's scent. Although lions are visual creatures, they undoubtedly have a far better sense of smell than human beings or anthropoid apes. I wondered what sort of response I would get from a wild animal with poor eyesight but very keen scent.

I had already tried this experiment with horses 25 years before – and, quite coincidentally, lent support to an artistic verdict given 2,300 years earlier by Alexander the Great. Having commissioned the famous painter Apelles to produce an equestrian portrait of him at Ephesus, he was dissatisfied with the artist's portrayal of his favourite charger, Bucephalus. In order to point out one or two errors, he had the horse led in front of the picture. Bucephalus whinnied as soon as he saw his likeness, whereupon Apelles smiled and remarked that the king's horse seemed to know more about painting than the king himself.

As it turned out I was delighted to find that Alexander was right and Apelles wrong. I could not, of course, proceed psychologist-fashion and simply ask my four-legged subjects for their opinion of a picture. I had to try to infer what they thought from their behaviour. My first step was to ascertain how horses react to unfamiliar members of their own species. As an officer serving with an army veterinary unit during the last war, I had ample scope for experiment. I began by introducing thirty-six horses – all hitherto unacquainted – to each other in pairs. It turned out that strangers in the horse world approach one another with heads held high and ears directed forwards. They then sniff one another, usually starting with the nostrils and proceeding to the tail and certain other regions of the body. What is more, they always stay close together in strange surroundings.

Page 344:
I took this portrait of a caracal or African lynx at liberty in the Ngorongoro Crater, Tanzania, not in a zoo. However, this particular caracal lived under canvas with us, enjoyed climbing on our laps, rode in the car, and went exploring in the neighbourhood. We had our work cut out to ensure that nothing happened to it.

When horse meets horse

Once I knew that, I introduced more than a hundred horses first to a life-size stuffed horse and then to various life-size pictures. The stuffed horse was greeted and treated in pretty much the same way as a live horse, and my subjects invariably remained standing near the dummy. If I drove a test-horse away from the manger by cracking my whip, it often vented its spleen on the defenceless stuffed horse. It would gallop over to the dummy, bite, kick, and sometimes knock it to the ground. After all, people who have been reprimanded by their superiors often take it out on their own subordinates or families – a form of behaviour popularly known in Germany as 'bicycling' (bowing to those up top and stamping on those below!). Even a crude life-size drawing of a horse on brown paper had its nose and tail sniffed. The mares and geldings could hardly be parted from it, and stallions tried to mount it. Completely stylized pictures, with simple columns for legs and angular outlines, were also treated to a certain extent as members of the same species. Apelles would have blushed to see what rudimentary pictures, well worthy of a modern artist, were fully acknowledged to be 'horses' by their own kind and treated as such. [A more detailed account of this appears in my book *Wir Tiere sind ja gar nicht so* (Franckh-Verlag, Stuttgart).]

These findings were all the more remarkable because horses possess large nostrils and huge expanses of olfactory mucous membrane – in other words, have a very much keener and more discriminating sense of smell than we do. On the other hand, as I was also able to prove by experiment, they are considerably inferior to us where sight is concerned.

But how would elephants, which presumably have still better scent and poorer sight, react to an inflated dummy elephant? For the past couple of weeks I had been the proud possessor of such a monster. I had written and telephoned around in Germany for almost two years without managing to persuade any manufacturer with a full order-book and a fat income to tailor me some inflatable giraffes, lions, elephants or rhinos out of plastic. In the end, the Nürnberger Gummi- und Plastikwaren-Fabrik took pity on

Trampled by a rhino

me. At the moment, the monstrous creature lay folded up small in a cardboard box behind my driving-seat like the bottled genie in Hauff's fantasy. I drove to see Ian Douglas-Hamilton.

Ian was a young British biologist who had been doing research into elephants in remote parts of the Manyara National Park, Tanzania, for the past two years. He once told me that it was one of my books which had prompted him to become a biologist. I was touched to hear that even my modest endeavours sometimes bore fruit.

Ian was accident-prone. Two years earlier a rhinoceros had trampled around on him just when his mother happened to be watching. The rhino ran off, but Ian could not walk and had to be helped back to his little house by his mother and a game warden. They then drove him to Arusha, where it was discovered that he had fractured a vertebra. The year before he had picked up bilharzia while bathing under a waterfall behind his lonely abode and was compelled to spend four months in England getting rid of it. Finally, and quite recently, a she-elephant had gored his brand-new Land-Rover, crumpling it and pushing it backwards ten yards into the bush.

Not that Ian had developed a fear of elephants. He simply pulled the inflated monster over his shoulders – in other words, walked along between the forelegs carrying the featherweight contraption on his back. And so the peculiar pale-coloured elephant made its way a trifle jerkily towards the group of real elephants whose shapes could occasionally be discerned in the surrounding bush. They were visibly interested in the new-comer, laid their ears flat, snorted, moved closer, and raised their tails. Then, one after the other, they turned and vanished into the greenery without a sound. I was meanwhile taking photographs, preferably so that the genuine and fake elephant would appear in the same picture. While I was at it a bull elephant unexpectedly emerged from the bush behind me. It is only on such occasions that one discovers how fast one can run.

Five times we approached groups of elephants, and each time

My grandson meets his first elephant

the same thing happened. The animals showed interest, approached and even threatened the stranger, then became suspicious and ran off. It occurred to us that this might be because our inflated elephant was much paler than its flesh-and-blood brothers. We sought out a mud-hole used for wallowing purposes by rhinos and Cape buffalo. The mud was coal-black and glutinous, and smelt extremely unpleasant. We anointed our plastic monster from trunk to tail, which at once made it look more authentic and elephant-like, but my right foot slipped and I went into the mud up to my calf. I ought to have stuck my left foot in as well, because my right shoe has never been the same shade since!

That evening, shortly before dusk, we drove to a remote part of the Manyara area seldom visited by tourists but often frequented by groups of elephants from outside the park. They are much warier than the regular inhabitants. Ian was driving yet another brand-new Land-Rover – a replacement for the one that had been wrecked by the she-elephant. Only the driver's cabin was enclosed, the loading platform at the rear being open. On this perched Stephan, the eleven-year-old son of my late son Michael. (It was his first visit to Africa and a very timely one – another year, and he would have to pay the full air-fare!) Beside Stephan squatted Alan Root.

Darkness was already falling when we saw a group of twenty-six elephants on a broad expanse of open ground. I was a little surprised when Ian carefully turned the car round, not a simple manœuvre on a narrow track hemmed in on both sides by boulders and undergrowth. As soon as the elephants spotted us they moved off in the opposite direction, all except three members of the herd, who advanced towards us. Elephants often move nearer in this way, but only for investigatory purposes. These animals spread their ears and raised their trunks to catch our scent, and one cow suddenly broke into a fast trot. Ian started up, on the better-safe-than-sorry principle. Elephants generally halt at a distance of ten or twenty yards, stand there for a while, and then retire.

Not altogether convinced

Not so this cow. She was in earnest. The car moved off as she bore down on us, but she was not content with driving us away – she meant to catch us. Ian regulated his speed so that she overhauled us slightly and then kept her distance. The speedometer registered 15 m.p.h. A charging elephant is an extremely unnerving sight, and Stephan was sitting on the open platform only ten yards ahead of this specimen. He banged on the rear window of the driver's cabin and asked to come and sit up front. I sympathized entirely, but we could not stop. The cow pursued us furiously for a good hundred yards, then pulled up as we accelerated. Stephan was highly impressed by our adventure.

During the next few days we repeatedly sought out small herds and lone elephants which Ian would approach with our mud-encrusted elephant-balloon on his shoulders. Two bulls showed keen interest and an evident desire to join battle. They came closer and closer, looking distinctly hostile, but their courage failed them. In the end they ran off at full speed, splashing through a stream which flowed into Lake Manyara. Not a single elephant consented even to touch our dummy with its trunk. We decided to try another series of experiments in the dry season. It was no use at present because the animals could find water everywhere.

Most men have a touch of ambition in them. We had been secretly impressed by the way in which Ian Douglas-Hamilton had boldly ventured up to wild elephants with the elephant-shaped balloon on his back while we were engaged in filming, photographing and recording the proceedings. When it came to rhinos, I felt an itch to swap roles. Everyone likes to prove to himself occasionally that he isn't a frightened rabbit. The rhinos in the Manyara district were too timid, the trees and undergrowth too dense and tall, so I drove back across the mountains to the Ngorongoro Crater and camped beside the Munge, a small river. Rhinos were always to be found on Ngorongoro's broad green expanses, which are devoid of undergrowth and entirely flat. Ngorongoro is the sixth-largest crater in the world, and its 100 square miles of floor is inhabited by more than 20,000 large animals.

Rhinos had no enemies but man

We had brought lots of blankets with us because the nights become extremely cold at 5,500 feet or more. Our tame caracal, a sand-yellow African lynx with handsome tufts adorning the tips of its ears, did not venture far from the tent. It did, however, fall into the fast-flowing Munge one night and had to be rescued by Alan Root. They both emerged with their teeth chattering. Wimpy, our tame cusimanse, was an insatiable explorer. She used to crawl up inside our shirts and appear at collar-level, burrow up our trouser-legs, scratch behind our ears and try to clean them. No scrap of dirt went unnoticed in any house inhabited by Wimpy, who would even rake it out from under the carpet. Excessive love of cleanliness can be irritating, so she was shut up in her box from time to time, especially after being found crawling between car-springs or examining the engine from inside.

During the late afternoon, night, and early morning, we were alone in the crater with the wild animals and the handful of Masai who lived at the far end. I drove round until I discovered a rhinoceros resting on a bare expanse of ground. It was asleep, and rhinos sleep soundly. They have had no enemies to fear for millions of years, nor have they yet adapted their behaviour to that late arrival on earth, man, let alone his diabolical invention the fire-arm. Rhinos can be extremely disgruntled if roused too abruptly.

One treats a rhinoceros with far more civility when walking around on two legs than when sitting in a car. I duly hailed the animal from a distance and, when that did not work, threw some stones near it. Before long its ears moved. It raised its head and rose leisurely to its feet. I walked slowly towards it carrying my inflated rhinoceros in front of me. The dummy was very light, consisting of nothing more than a thin plastic skin filled with air. 60 or 70 yards away stood the car, where the others were sitting with their telephoto lenses and binoculars. I continued to advance slowly on the lone rhinoceros, crouching below the hind quarters of my dummy so that the genuine article would not spot me.

Rhinos have very poor sight – that much I knew. A bull rhinoceros trailing a cow on open ground which permits the human

Then we tried rhinos

observer a clear view of both animals does not head straight for the object of his desires but laboriously and circuitously traces her route by scent alone. That is why rhinos attack so readily. They cannot see what confronts them, so they charge their adversary on principle but often stop short or swerve at the last moment, as we have already mentioned. On the other hand, it is never easy to tell precisely what a rhinoceros will do. Only a few weeks before, a five-ton truck had been attacked by a rhinoceros as it returned from the Wogakuria game warden's post, here in the Serengeti National Park. One of the front tyres was punctured and the wing buckled.

My bull rhinoceros slowly advanced, growing more and more agitated. His little tail went up. He snorted, raised his head, lowered it again, trotted a few paces towards me, or, rather, my rhino-shaped balloon, retreated once more, and started to frisk about just as bull rhinos do when they encounter each other for the first time. Each hopes to intimidate the other, persuade him to turn and run away. I did not oblige my opponent. He grew bolder, but as soon as I raised my inflated rhinoceros and waggled it slightly his courage receded and he withdrew again.

I entirely forgot that only air and a thin film of plastic separated me from the aggressive rhinoceros which was my only companion for some distance around. I straightened up because it was too uncomfortable to sustain a crouch any longer and put on a hat to stop my nose catching the sun. I felt like a torero in the arena. Both I and the bull rhinoceros seemed to be enjoying our game. We capered around, jockeying for position. The bull did not appear to notice that his opponent had no rhino aroma, even though rhinos smell quite strong as a rule. He was far too excited and resentful of my presence. On the other hand, he did not venture to attack in earnest because I showed no signs of intimidation. His horn only once brushed the head of the dummy. I was afraid he would notice how soft it was, but he did not. If he had really gored my thin-skinned rhino I should have been left standing there on my own, but I calculated that he would be so taken aback that I

Safety in the VW

should have time to run for the car. Fear lends a man wings, as I knew from similar experiences in the past. Also, Alan Root and the rest of the party would no doubt race to meet me.

But nothing happened. My opponent did not pluck up enough courage to attack, and so, when things began to get boring, I retreated towards the Volkswagen bus. I had to walk backwards with the head of my dummy facing the real rhino. If I had turned round, he might have taken it as a sign of fear and defeat and charged me from behind. As it was, he first went to the spot where I had just been standing and sniffed the ground with great interest. Then, still sniffing the ground, he started to trail me. Eventually he halted and released a veritable cascade of urine, plumb on my track. As for us, we drove away. We had many more adventures with rhinos in the days that followed, some of which are recorded in our photographs.

Bibliography

A few important and more detailed works

ALLAN C. BROOKS and IRVIN O. BUSS: 'Past and present state of the elephant in Uganda,' *Journal of Wildlife Management*, vol. 26, no. 1, 38-50, 1962.
IRVIN O. BUSS and NORMAN S. SMITH: 'Reproduction and breeding behaviour of the African elephant,' *Journal of Wildlife Management*, vol. 30 (2), 275-388, 1966.
H. B. COTT: 'Life of the Nile crocodile,' *Black Lechwe*, vol. 3, no. 3, 4-13, 1962.
JOHN GODDARD: 'Food Preferences of two black rhino populations,' *East African Wildlife Journal*, vol. 6, 1-18, 1968.
C. A. W. GUGGISBERG: *S.O.S. Rhino*, André Deutsch, London, 1966.
C. A. W. GUGGISBERG: *Simba, eine Löwenmonographie*, Hallwag, Berne, 1960.
HANS HIMMELHEBER: 'Die Geister und ihre irdischen Verkörperungen als Grundvorstellung in der Religion der Dan (Liberia und Elfenbeinküste),' *Baseler Archiv*, New Series, vol. 12, 1-88, 1964.
A. MARIA HOYT: *Toto and I. A gorilla in the family*, Lippincott, Philadelphia, 1941.
HANS and UTE KLINGEL: 'The rhinoceroses of Ngorongoro Crater,' *Oryx*, vol. 8, 302-306, 1966.
WOLFDIETRICH KÜHME: 'Beobachtungen an afrikanischen Elefanten in Gefangenschaft,' Part 1: *Zeitschrift für Tierpyschologie*, vol. 18, 285-296, 1961. Part 2: *id.* vol. 20, 79-88, 1963.
HUGH F. LAMPREY and MYLES TURNER: 'Invasion of the Serengeti National Park by elephant,' *East African Wildlife Journal*, vol. 5, 151-166, 1967.
JANE VAN LAWICK-GOODALL: 'My life among wild chimpanzees,' *National Geographic Magazine*, Washington, 272-308, 1967.
R. M. LAWS: 'Eye lens weight and age in African elephants,' *East African Wildlife Journal*, vol. 5, 46-52, 1967.
R. M. LAWS: 'Age criteria for the African elephant,' *East African Wildlife Journal*, vol. 4, 1-55, 1966.
R. M. LAWS and J. S. C. PARKER: 'Recent studies on elephant populations in East Africa,' *Symp. Zool. Soc. London*, no. 21, 319-59, 1968.

M. L. MODHA: 'The ecology of the Nile crocodile,' *East African Wildlife Journal*, vol. 5, 96-105, 1967, and vol. 6, 81-88, 1968.

A. C. POOLEY: 'The Nile crocodile, *Crocodilus niloticus*,' *Lammergeyer*, vol. 2, 1-55, 1962.

A. C. POOLEY: 'Preliminary Studies on the Breeding of the Nile Crocodile in Zululand,' *Lammergeyer*, no. 10, 22-59, 1969.

A. C. POOLEY: 'Burrowing Behaviour of Crocodiles,' *Lammergeyer*, no. 10, 60-63, 1969.

VERNON REYNOLDS: *Budongo. A Forest and its chimpanzees*, Richard Clay Ltd., Bungay, Suffolk, 1965.

A. T. A. RITCHIE: 'The black rhinoceros,' *East African Wildlife Journal*, vol. 1, 54-62, 1963.

E. FRANZ SAUER and ELEONORE SAUER: 'Verhaltensforschung an wilden Straussen in Südwestafrika,' Survey in *Wissenschaft und Technik*, vol. 67, 652-657, 1968.

GEORGE B. SCHALLER: *The mountain gorilla*, University of Chicago Press, 1963, id. 'Studies on lions in the Serengeti,' *National Geographic Magazine*, Washington, April, 1959.

RUDOLF SCHENKEL: 'Zum Problem des Territoriums und des Markierens bei Säugern – am Beispiel des Schwarzen Nashorns und des Löwen,' *Zeitschrift für Tierpsychologie*, vol. 23, 593-676, 1966.

RUDOLF SCHENKEL and DR LOTTE SCHENKEL-HULLIGER: *Ecology and behaviour of the Black Rhinoceros. A field study. Mammalia depicta*, Paul Parey, Hamburg, 1969.

SCHMIDT-NIELSEN and T. R. HAUPT: 'Thirst of dromedaries,' *Scientific American*, Dec. 1959, 140.

R. V. SHORT: 'Oestrous behaviour, Ovulation and the formation of the corpus luteum in the African elephant,' *East African Wildlife Journal*, vol. 4, 56-68, 1968.

S. K. SIKES: 'The elephant problem in Africa,' *African Wildlife*, vol. 20, no. 3, 225-237, 1966.

SYLVIA K. SIKES: 'Habitat stress and arterial disease in elephants,' *Oryx*, vol. 9, no. 4, 286-292, 1968.

C. A. SPINAGE: *The Book of the Giraffe*, Collins, London, 1968.

Index

aardvark 325, 332
Abdullah 233–6
Aberdare Mountains 266
Abren, Mrs 237
acacia plant 56
   tree 266
Achard, Peter (game warden) 12–13, 15
Acholi 276
Acokanthera Schimperi 62
Actitis hypoleucos 163
Adamson, Joy 70, 72, 121, 174, 178
Adamson, George 174
Addo National Park 277, 295
Aelian 163
Africans 15, 29, 73, 94, 97, 104, 109, 113, 144, 164, 167, 172, 197, 200, 201
African eagle owl 173
Afro-Asian Conference 208
Albert, Lake 111, 155, 170
Alexander the Great 345
Algeria 103, 126
Alice 212
alligator 93, 109, 156
   American 177
Amazonian jungle 77
Amboseli Game Reserve 39, 40, 49, 60, 63–4, 65, 72, 76, 131, 141
Amphicar 111f, 190
anæsthetic darts 65, 132
Angola 42, 43, 220, 319
Antarctic 187
antelope 17, 30, 56, 69, 72, 74, 89, 90, 106, 147, 150, 151, 178, 181, 182, 197, 282, 285, 300, 326, 327
   blue-buck 319
   Bubal hartebeest 182
   dik-dik, Guenther's 80
   duiker 229
   dwarf 20
   eland 89, 150, 268; Cape eland 197
   gazelle 69, 136, 141, 142, 324
   gemsbock 121
   gnu 35, 49, 51, 73, 75, 132, 133, 134, 136, 141, 142, 181, 268, 326, 328–9, 334, 336
   Grant's gazelle 325–6
   impala 167, 282, 324, 326, 327, 334
   kudu 44, 89, 268, 273, 327–8
   oryx 143, 197
   reedbuck 203, 328
   roan 35, 328
   roe deer 89
   springbok 49
   Thomson's gazelle 17, 35, 75, 133, 141, 325–6, 328, 331–2
   topi 89
   Uganda kob 89–90
   waterbuck 54, 89, 282, 285, 300, 326, 327
ant-heap 30
Anti-Atlas Mountains 125
Antonius, Professor 70
Antwerp 13, 16, 24
ape, anthropoid 11, 14, 16, 17, 20, 23–4, 30, 35, 209, 211, 223, 225, 226, 231, 337, 345
Apelles 345, 346
Arabia 102, 255
Arabs 100, 200–5
'Archie' 233
Aristotle 163
Arusha 14, 15, 32, 35, 70, 161, 201, 333, 347
   National Park 53–4, 148
Asia 38, 144, 181, 260, 261
Assi 337–8
Atilax paulinosus 173
Atlas Mountains 125, 256
Attwell, R. J. G. 165, 326
Augustus, Emperor 177
auroch 68,
Australia 183–4, 261, 320

Baard (game warden) 311–12
baboon 21, 148–9, 220, 221
Backhaus, Dr D. 153, 318
Baker, Mrs 110
Baker, Sir Samuel 110, 265

Index

Balestra, F. 143
bamboo 299
Bambu 11
Banks, Sir Joseph 188
baobab tree 108, 266
Barcelona Zoo 220
Barnum and Bailey's Circus 185
Barotseland 319
Basle Zoo 64, 218, 288, 309, 312–13
Batawan tribe 156
Bauer, Mr and Mrs 279
bear 68, 150, 178
Bechuanaland *see* Botswana
bedouin 259
Beebe, William 186
Belgian game department 265
Belgians 210, 262
Belgian Congo 321
Belgium 185, 210
Beni Abbès 256–70
Bere, R. M. 313
Berggreen Zoo Park of Sweden 13
Berlin 11
 Zoo 64, 134
Berne Zoo 178
Besser, Hans 164–5, 171–2, 177
bilharzia 24, 32, 108, 347
Billy 275–6
bird 89, 174, 178, 180, 185, 197, 229, 300
bird of paradise 68
bison 68
blackbird 338
Black Sea 124
Blagden, Dr Charles 187–9
Blake, Alexander 164
Blanquita 240
bluebuck *see* antelope
bongos 18
bonobos 14
Borneo 93
bot-fly 55
Botswana (formerly Bechuanaland) 123, 156, 319
Brain, C. K. 144–5
Brandenburg 124
Brehms Tierleben 130
Breslau 125
 Zoo 331
Brisbane 183
Bristol Zoo 79
Britain and the British 13, 43, 93, 96, 124
Bronx Zoo 331
Brunswick 186

Brydon, Mr 316
Bubal hartebeest *see* antelope
Bucephalus 345
budgerigar 338
Budongo Forest 20, 23
Buechner, H. K. 300
buffalo 35, 54, 56, 165, 182, 229, 314, 335
 Cape 52, 54, 72, 74, 110, 114, 148, 161, 165, 166, 178, 197, 282, 284–5, 316, 348
Bukindo 17
Bundestag, German 208
Burchell, William 319
Burhinus vermiculatus 163
Buss, Irvin O. 288, 298, 302, 303
bustard 178
Buxton, Aubrey 121
buzzard 178

cabbage, Nile 110
Cade, C. E. 329–31
California 181
Cameroons, French 42, 230
camel 9, 106, 150, 185, 254f, 260
 caravan 254
 dromedary 107, 191, 197, 254f
 food and drink 254–5
 physique 254f
 saddle 254–5
 speed 254
Canada 181
Cannon (big-game hunter), 42–3
Cape buffalo *see* buffalo, Cape
Cape eland *see* antelope
Cape Province 319
Cape Times 279
Cape Town 71, 262
caracal *see* lynx
Carthaginians 262
Caspian Sea 17, 124
Castro, Fidel 235
catfish 172
cats 240, 246
cattle egrets 55
cave-drawings 319
cave-lion 67–8
Central Island (Lake Rudolf) 161–2, 169–70
Ceratotherium simum cottoni 321
Ceratotherium simum simum 321
Chabert, Ivan Ivanitz 187
chameleon 190
Chapman, Prof. David 122
Chapman, R. 128
Charly 275

Index

cheetah 74, 133, 284
Chicago Zoo 59, 65
chicken 99, 134, 329
Chikwawa 168
Chilongori reservation 327-8
chimpanzee 11, 13f, 30f, 221, 225, 229
  and water 20, 22, 23, 32-3
  attitude to young 22, 221
  'carnival' 21-2
  communication 21-2, 32
  curiosity 30
  dancing 21-2
  environment 20
  food 20-1, 23, 225
  greeting/friendship 21, 22, 30, 221
  grouping 21
  intelligence 30-1, 226
  in Zoo 13-14, 16, 19
  maturity 11
  medical use 17-18
  nests 33
  on Rubondo *see* Rubondo
  'Operation Chimpanzee' 12-35
  release 19-23, 24, 36
  shipping 13-14
  study of 19, 20, 21-2, 23, 36
  violence 11-12, 19, 36-7, 225
  West African 20
  with humans 14, 17, 19, 22, 23, 30-2, 36
China 40, 68, 93, 205
chinchilla 103
Chingola 108
Chipandale, Elard 168
Chirundu 294
Chobe river 42
Christians 97
Chulka river 38
Circuses 177, 185
  Barnum and Bailey 185
  Ringling Brothers' 246-7
  Sarrasani 185
Cleveland 185
cockerel 233
Columbus Zoo 223
Congo 42, 57, 112, 113, 179, 205, 209, 210, 233, 247, 262, 265, 266, 279, 280, 281-2, 286, 293, 309, 311, 318, 321, 337
Congo Basin 42
Constance, Lake 192
Copenhagen Zoo 13
cormorant 34, 192
Cotonou 97
Cott, Hugh B. 111, 161, 164, 170

cow 39, 95, 97, 255; cattle 89, 167, 186, 191, 210, 317
Cowie, Mervyn 58, 133-4, 143
Cox, William 108-9
crabs 174
Cracow Museum 38
Creation, The 96
cricket 174
Critchley, R. A. 328, 332
crocodile 9, 18, 32, 56, 93, 108-11, 155f, 192, 198, 283, 327
  American 177
  and birds 162-3
  as exhibits 177
  environment 155-6, 161-2
  food 110-11, 165-6, 174
  in Zoo 111
  locomotion 163-4
  mating and young, 169-76
  physique 110, 162-3
  preservation 155
  religious significance 108, 167, 168
  skin 18, 109, 156, 161
  violence 108-9, 161, 164-5, 166-8, 171
Cros de Cagne Zoo 134
crow 180
crowned cranes 336
Cuba 232, 235, 242
cusimanse 350
Cyprus 125

Dahomey 96-7
Dan 167, 168, 314-15
Dar es Salaam 15-16
Darling, F. Fraser 50
Darwin, Mount 143
dassies 145, 261
  tree 229
Datschitz 338
Davison, E. 266
de Clerck (game warden) 153
deer 60, 141, 179
De La Bat (game warden) 149
Denmark 67
Diceros bicornis 39
Diceros bicornis somaliensis 39
diospyros-fruit 295
dikkop, water 163
diver 186
dog 73, 135, 179-80, 285, 333, 337-8
  bull terrier 41
  dachshund 285
  fox terrier 55

357

Index

dog [contd.]
  gun 333
  hound 333
  see also wild dog
donkey 255, 258
Don Steppe 125
Douglas-Hamilton, Ian 347-52
dove 134, 225
dragon-fly 174
dragons' teeth 40
Dresden Zoo 69, 80
dromedary see camel
duiker see antelope
Dumacheri 17
dummies see elephant, giraffe, horse, lion, rhinoceros, zebra
Dunnett, Sinclair (game warden) 15, 16, 24, 29-34
Dürer 40
Dustbin Nelly 275
Dutch 132, 262
Dutch East Indies 92
Dyak 93
dwarf antelope see antelope

eagle 178
East Africa 16, 18, 42, 43, 57, 73, 74-5, 164, 182, 190, 191, 199, 206, 232, 263, 265, 267, 298, 320, 326
  lakes 192
  mountains 42
  national parks 18
Ecuador 92
Edward, Lake 112-13, 155, 264, 334
Edy (game warden) 60
Eglinton 316
Egypt 102, 124, 167, 205, 335
Egyptians, Ancient 102, 144, 167, 335
Eibe-Eibesfeldt 338
Eibe Oldendorff 13, 15
Elagabalus, Emperor 177
eland see antelope
Eldoret 147
elephant 9, 18, 38, 43, 44-51, 64-5, 67, 74, 110, 113-14, 116, 143, 165, 178, 182, 184, 185, 197, 229, 248, 261f, 316-17, 334, 335
  African 262f, 316; bush 263, 264, 287, 294, 298; forest 263, 264
  Asiatic or Indian 262f, 316
  behaviour 281, 282-3, 290-1
  circus 282, 285, 286, 294
  Congolese 313
  controlled shooting of 301, 302
  dummies 304, 309, 346-9
  enclosing 276-8, 286, 290
  environment 267, 273-4
  excretion 51, 298
  food 266, 267, 273, 275-6, 297, 298-300
  graveyards 293-4
  grouping 288-90, 310, 313
  hunted 43, 264-5, 266, 298
  investigation—Ford Foundation subsidised 273
  in water 287
  in zoos 261, 282, 285, 286, 287, 294, 296, 297
  ivory market 262, 265, 293, 302
  mammoths 261
  'musth' 302
  North African 262
  numbers 261, 266-7, 273, 281, 300, 301-2
  physique, speed 262-4, 285-6, 294-7, 313
  Pygmy 264
  relationship with: hippopotamus 283; other animals 282-4, 300, 314; rhinoceros 53-4, 283, 300
  religious significance 314-15
  sleep 286-7
  subungulate 261
  trained 262, 282, 296, 297, 309
  tusks 261, 262, 263, 264-5, 287, 293, 294-5, 303
  violence 266, 277-81, 283-5, 291-2, 301
  voice 287-8
elephas africanus pumilio 264
Elizabethpol 125
elk 68, 179
Ellis (game warden) 52
Eloff, F. C. 133
Ellen 212
Elsa 69, 70, 72, 174, 178
Emin Pasha 320
England 15, 67, 122, 129, 185, 186, 347
Entebbe 113, 190, 200, 202, 208
Ephesus 345
Epirus 262
Equatoria 200, 205
Estes, R. 329
Ethiopia 42, 197, 263
Etosha National Park 44, 49, 80, 136, 248, 311
Etosha Pan 44, 49, 149, 283
euphorbia-tree 50, 115
Europe 11, 16, 17, 36, 38, 89, 134, 141, 163, 179, 181, 185, 187, 247, 261, 320
European Zoos 13, 15, 36, 168

Fajao 313

358

Index

fat-headed fly 65
Faure, M. 129
Faust, Dr Ingrid 101
Faust, Dr Richard 101
Felicia 56
filarial worms 65
fire 178–187, 293, 337
  animals' reaction to 178–9
  as protection 178
  bush 50, 178, 181, 184, 267, 268, 297
  circus 185
  cleansing by 186
  explosives 186
  fascination of 179–80
  forest 181, 183–4
  man made 182–4, 273
  snake farm 186
  spread of disease as a result of 181–2
  storms 179
  volcanic eruption 186
  zoo 185
fish 89, 110, 163, 166, 174, 338
  vundu 165
Fitzsimmons, F. W. 184
flamingo 116
Flamian Circus 177
Florida 99, 103, 109, 156, 247
Ford Foundation 273
Fort Archimbault 42
fox 141
  silver 103
framboesia 144
France 67, 124
  south of 103
francolin *see* partridge
Frankfurt 14, 35, 112, 208, 290
  Zoo 12, 14, 17, 36, 39, 51–2, 57, 58, 59, 88, 90, 91, 92, 94, 95, 96, 99–100, 101, 116, 150, 153, 179, 189, 207, 218, 225, 297, 318–19
Frankfurt Zoological Society 13, 121
French African Colonies 42
French Cameroons 42
French Guinea 19–21
French West Africa 232
Fritz Thyssen Foundation 274
frogs 80, 89, 174, 178
Furuviks Parken 13

Galana River 267
Galapagos Islands 186–7
Galicia 124
Gandersheim 186

Ganga na Bodio 262, 263, 282, 286, 287, 291, 297–8
Garamba National Park 44, 153, 318, 321
Gargantua 246
Gebbing, Herbert 51–2
Geese, laughing 185
Genet, rusty-spotted 173
Genetta rubiginosa 173
German East Africa *see* Tanzania
Germany 67, 124, 185, 313, 317, 346
Gerstenmaier, Dr 208
Gerty 39, 41, 60, 65
Gevers, Prof. 279
gibbon 69
Gibbons, Major A. 320
Gillett, C. 279–80
giraffe 9, 35, 72, 106, 136, 146f, 156, 182, 185, 192, 321, 335
  as prey 149
  attacks by 147–8, 149, 150
  birth 152–3
  colouring 153–4
  dummies 346
  enclosing 147
  environment 148–9
  grouping 149–51, 152
  in Menengai Crater 147
  in Rubondo 35
  in zoos 150, 152, 153, 154
  physique 146–7
  superstition 154
  vision 153
Gir Forest 71
goat 35, 88, 91, 106–7, 133, 147, 191
  Indian long-eared 90
Goddard, John 49, 58, 329, 332
Gombe Forest 20, 21–2
  National Park 225
  Reserve 20
Goodall, Jane 19, 20, 21–2, 23, 225
Goose, African spur-winged 80
Gordon, A. 300
gorilla 9, 14, 18, 19, 20, 209f, 217, 218
  baby 14, 218
  behaviour 221–2, 226–7, 229, 231
  capture 233
  communication 222
  environment 210–11, 228
  evil image 209, 231
  food 20, 219, 225–7
  grouping 211–12, 217–18, 220–2, 230
  intelligence 226, 235f
  in water 20, 226, 227

Index

gorilla [contd.]
  in zoo 14, 218, 223, 225, 226
  lowland 232
  mating, pregnancy and young 217-20, 221, 222, 233f
  mountain 209-11, 217, 227, 232, 244
  nests 227, 228
  non-violence 217, 226, 227, 229-30, 231
  physique 220
  play 221, 235f
  stare 222-4
  territory 217
  violence 222-3, 230
Gotha 125
Gottfried Fridh 13
gnu 51, 73, 75, 132, 133, 134, 136, 141, 142, 181, 268, 326, 328-9, 334, 336
grass 182
  magugu 299
Grant's gazelle see antelope
Greece 44, 67, 183
grouse 179
Grzimek, Michael 19, 21, 49, 85, 204, 268, 278, 282, 348; Michael Grzimek Memorial Laboratory 274, 282
Grzimek, Stephan 348-9
Guggisberg, C. A. W. 54, 56, 73, 163
guerenza monkey 35
Gyrstigma conjugens 65
Gyrostigma pavesii 65

Hagenbeck, Carl 106
Halfway House 94
Hannibal 262
Hanover Zoo 52, 59, 99
hare 141, 145
Harris, Cornwallis 316
hartebeest, Bubal see antelope
Harvey, Gordon (game warden) 16
Haupt, T. R. 256
Havana 232, 235, 240
hawk 172
Hay, P. (game warden) 90
heat experiments 187-9, 257-9
Heck, Ludwig 130
Hediger, Prof. 286
Heidelberg 168
Heilbron 152
hen 88, 232-3
Henderson, J. 185-6
Heppes, J. B. 317
Herodotus 127, 162-3
heron 172

Hessen 67
Heterobranchus longifilis 165
Himmelhever, Dr Hans 167-8, 314
Hipoplopterus spinosur 163
hippopotamus 24, 36, 56, 110, 112-13, 116, 165, 192, 207, 282, 327, 334
  violence 24, 36, 56
Hluhluwe Game Reserve 62-3, 64, 320, 322
Höhnel, Lieutenant 190
Hoier, Col. R. 293, 311
horse 30, 55, 74, 76, 150, 254, 255, 285, 297, 317, 329, 345-6
  wild 74
Hoyt, E. Kenneth 232-43
Hoyt, Maria 232-53
Hunter, John A. 40, 43, 60
hunters 13, 40, 42-3, 60, 62, 63, 67, 70, 73, 75-6, 93, 97, 101, 106, 130, 152, 156, 164, 197, 230, 231, 233, 261, 262, 264-5, 288, 293, 315, 317, 324, 333
Husain 129
hyena 9, 49, 56, 69, 74, 104, 290, 327, 328, 334-5, 340
  as hunters 133-4, 136, 141, 142-3, 144
  as scavengers 133
  brown 144
  carrion eaters 130
  cave 134
  food 130
  grouping 131, 132
  in zoos 134-5
  observation 132-3
  physique 135
  relationship with other animals 74, 130-1, 144
  spotted 132, 133, 134, 135-6, 141, 142-5, 334-5
  striped 144
  superstition 144
  taming 130, 135-6, 144
  young 135
Hypanhenia 182

ibex 39
Iceland 187-8
Idunda 312-13
iguana 93
ikitos 201, 205
Immelmann, Dr Klaus 100
impala see antelope
Indefatigable Island 187
India 64, 71, 78, 97

Index

Indian Ocean 161
Indo China 232
Indricothericum asiaticum 38
Innes, Mr 152
Ionides, C. 328
Iraq 102
iron-wood tree 21
Israel 155, 183
Israeli 183, 258
Italian 198
Italy 18, 183, 262
Ituri Forest 282
Ivory Coast 16, 167, 232, 233, 274, 286, 314

jackal 74, 130, 133, 142
jackdaw 180
jaguar 68
Japan 232
Jarnum, Dr 256
Jimmy 36-7
Johannesburg 94
Johnson, Martin 57, 61
Johnson, Mrs 61
John the Baptist 127
Jordan 102
José 243, 245
Juba 202, 204, 207

Kabara 211, 219, 227, 229
Kabue 108
Kachari 313
Kade, Ulrich (game warden) 34, 36
Kalahari Desert 121, 183
  National Park 133, 143
Kampala 190, 200
Karamoja 148
Kariba Dam 41, 90, 276
  Lake 41-2
Kassala 107
Katete, Francis 121
Katete River 289
Katherina die Grosse 59
Katwe 334
Kayama 242
Kayonza 230
Kazakhstan 38
Kazinga Channel 12
Kazuga inlet 287
Kearton, Cherry 55, 63, 144
Kenya 14, 18, 39, 40, 42, 43, 51, 61-2, 65, 72, 112, 133, 147, 154, 161, 168, 183, 190, 198, 206, 254, 265, 266, 267, 273, 281, 290, 320, 327

Game Department 198
Game Reserve 147
  Mount 67
Khartoum 205, 206, 208
  Zoo 207
Kibo 334
Kidepo National Park 199
kidney disease 18, 108
Kilimanjaro 76, 136, 154, 334
Kilombero 283
Kilwa 282
Kirawira Plain 156
Kirby, F. Vaughan 319-20
Kisoro 226, 227, 229
Kitsombiro mission hospital 230
Kittenberger, Kalman 59
Kiwu 244, 293
Klingel, Hans 49, 50-1
Klöppel, Dr 59
Klose, Horst 12, 36
koala bear 184
Koenig, Oscar 63
Kolb, Dr 43
Kollmannsperger, Franz 127
Kolozi River 313
Kotlandt, Dr Adrian 20, 24
Krakatoa 85
Kreth 168
Krieg, Prof. Hans 68
Kruger National Park 54, 70, 73, 141, 144, 151, 152-3, 166, 263, 266, 279, 283, 285, 296, 297, 299, 320, 324
Kruuk, Dr Hans 132-3, 135-6, 141, 331-2
Kruuk, Mrs 132-3, 135
kudu see antelope
Kühme, Dr Wolf Dietrich 282, 290
Kyle Dam Reservation 320

Land-Rover 55
Langley-Elton, Mr Robert 123
lava 186
Lawick-Goodall, Jane Baroness van see
  Goodall, Jane
Laws, Dr 294, 295
Lederer, Dr Gustav 87, 91
leeches 163
Leiden Museum 71
Leningrad, Physiological Institute of 146
leopard 11, 35, 68, 70, 85, 90, 94, 133, 135, 136, 147, 229, 326
  black 229
Lerai Forest 55-6
Letaba camp 285

361

Index

Lewin, A. 304
Liberia 167, 314
Liège 185
Limpopo 316
Linnaeus 187
lion 9, 18, 35, 44, 49, 61, 67–76, 110, 115, 130, 136, 283–4, 309, 336–7, 338–45
   and fire 178
   Berber 71
   Cape 71
   cave-lion 67–8
   dummies 304, 336, 338f
   effect on humans 67–8
   family life, mating, breeding, etc. 69–70
   food 74–5, 326
   hunted 75–6
   hunting 72–3, 74–5, 76, 115–16, 121
   in trees 70
   in zoos 70, 74, 189
   physique, leaping, laziness, and speed 68, 70–1, 74
   relationship with other animals 55–6, 67, 71–2, 133, 149, 283–4
   roar 68–9
   status of 6, 7, 115
   steppe-dwelling 68
   white 70
Litou-Kiperere 299
Livingstone, David 73, 155, 265–6
lizard 80, 178
   monitor 90
   Nile 90
   Rock 90
locust 124f
   as food 127
   environment 128
   food 127
   in Bible 124
   mating, breeding and development 126–9
   migration 125–6, 228–9
   migratory 128
   plagues 124–5
   Red Migratory 128
Locusta Migratoria 125
Locustana pardalina 125
London 71, 123, 180, 187
   Natural History Museum 71
   Zoo 79, 150
Longwoo 184
Lord Mayor 275
Lorongsa, Kartua (game warden) 328
Loudetia superba 182
Loxodonta africana 263

Loxodonta cyclotis 263
Luamfia Lagoon 327
Luangwa River 43, 165, 166
Luangwa South Game Reserve 295
Lubero 230
Lulua River 179
Luxemburg 67
lynx, sand-yellow African (caracal) 345, 350

Machadodorp 94
Madagascar 161, 171
magugu grass 299
Mahdi rebellion 321
Makula 218
Makumba 282
malaria 20
Malawi (formerly Nyasaland) 43, 142, 168, 276
Malaya 97
Malelane Road 153
Malinda (game warden) 60
Malomi 109
Manchester Zoo 185
marula-tree 283, 299
Manyara National Park 55, 60, 70, 278, 304, 309, 347–52
Manyara Lake 304, 349
Marais, J. J. 95
marabou stork 163, 172, 178
Marshalls Creek 186
Mars Ultor, temple of 177
Martin, P. 90
Masai 40, 62, 350
Mashi River 319
Masindi Hospital 280–1
Matapos National Park 320
Mathur 129
Max 14, 212, 218, 253
Maxwell, John 108–9
May, Karl 254
Mbagathy 133
Mbarara district 328
Meinertzhagen, Col. 41
Menengai Crater 147, 150
Meru, Mount 97, 324, 333
Mez, Dr Theodor 206, 208
Michael Grzimek Memorial Laboratory 274, 282
Mikumi National Park 131, 327, 333, 335
mink 103
Mitchell, B. L. 182
Mkomazi Game Reserve 136, 154
Mkuzi Game Reserve 171, 174–5, 248

Index

Mlanje 142
Modha, M. L. 161, 170, 173
Molloy, Mr 49
Mombasa 14, 15, 16, 112, 267
Momella (farm) 333
mongoose, marsh 173
　dwarf 248
monkey 68, 96, 185, 197, 225, 237
　howling 69
　guerenza 35
Morocco 125, 129, 232
Moroto 148
Moshi 15, 63, 64
mosquitoes 20, 229
Mountain Gorilla, The (Schaller) 209
Mozambique 42, 43, 263
mtomboti-tree 50
Mtuma 164-5
mugongo fruit 299
Munge River 349-50
Munich Zoo 262, 313
Murchison Falls National Park 54, 109-10, 111, 116, 153, 155, 170, 265, 275, 278, 280, 283, 286-7, 293, 294, 298, 299, 300, 322, 323
Murchison Bay 167
mussel shells 164
Mutesa, King 167
Mwanza 12-13, 15, 16, 17, 34, 35
Mweya Lodge 134, 287
Mzima Springs 56, 168

nagana 181
Nairobi 18, 122, 133, 147, 206, 267
　National Park 52, 54, 101, 330
　Zoo 330
Namsika 287
Namutoni, Fort 136
Nana 92
Napo River 92
Natal 50, 54, 62-3, 167, 175, 319
National Parks and Reserves 109, 112, 302-15, 320, 324; *and see:* Addo, Amboseli, Arusha, Chilongori, Etosha, Garamba, Gombe, Hluhluwe, Kalahari, Kidepo, Kruger, Kyle Dam, Luangwa, Manyara, Matapos, Mikumi, Mkomasi, Mkusi, Murchison Falls, Nairobi, Ngorongoro, Nimule, Queen Elizabeth, Parc National Albert, Rukwa, Serengeti, Toro, Tsavo, Umfolozi, Wankie
Nemsi, Karaben 254
Nepal 97

Nero 179-80
New Guinea 156
newspaper 15, 16, 206, 266, 273, 279
New York Natural History Museum 233
New York Zoo 79, 223, 264
New York, Zoological Society of 78
Ngoma 94
Ngongongare (farm) 333
Ngorongoro Crater 12, 18, 40, 41, 49, 55-6, 58, 71, 116, 132, 136, 141, 143, 304, 324-6, 329, 332, 335, 336, 345, 349
Ng'uni, Patulan 289
Ngurdoto Crater 61, 148
Nichols, Mr 316
Nicholson, W. D. 283, 288-9, 299
Niger 129
Nigeria 42, 96, 232
Nile 90, 110, 121, 161, 202, 204, 206, 283, 313, 316, 323
　Semliki 155, 176
　Upper 320-1
　Victoria 108, 110, 111, 155, 163, 169, 287, 301
Nile Cabbage 110
Nile varan 170, 172-3
Nimba Mountains 19, 21
Nimule National Park 316-17, 321
Nko Forest 220
Noack 264
Nomadacris septemfasciata 125
Nomadacris septemfasciata Serville 128
No room for wild animals (Grzimek) 262, 282
North Africa 102, 144, 183, 254, 255, 260
Northern Rhodesia *see* Zambia
Norway 67
Nuanetsi district 304
Nuremberg Zoo 286
Nurnberger Gummi- und Plastikwaren-Fabrik 346-7
nutria 103
nyakahimbe 21
Nyakyusa 225
Nyamugasani River 280
Nyasaland *see* Malawi
Nyerere, Dr Julius 10, 208
Nzega 295

Obaha, Daniel (game warden) 37
Obongi 322-3
Ochara, N. L. (game warden) 144
okapis 18, 282
Okavango Marsh 156
　Swamp 123
Olduvai Gorge 50

363

Index

Ol Tukai (tourist camp) 72
Omar, Hadji Halef 254
Opel Zoo 282, 290, 313
'Operation Chimpanzee' 11-37
Orange Free State 125
Orange River 319
orang-outang 14, 20, 225
   Asiatic 225
   in Zoo 14
oryx *see* antelope
ostrich 80, 99f
   aggression 99-100
   braveness 104
   breeding, eggs and young 99, 101, 102-4
   control of 99
   environment 104-5
   farms 80, 99, 102-3
   feathers 99, 102, 103, 106
   food 104-5, 106
   grouping 105
   in zoos 99-101, 106
   legends 100
   physique, speed 99-100, 103-4
   riding 100
   sleep 100
   young 103
Otto 150, 153
Ouidah 97
Owen, John 61, 201, 203, 204, 274
Owen, R. 209
owl, African eagle 173

Pacific 187
Pakuba 322
Palestine 67, 155
Paraa Lodge 275, 287, 299
Parc National Albert 210, 264, 279, 328
Parker, Ian 109
Paris Zoo 79, 95
partridge 339
passion fruit 299
Pectropterus gambiensis 80
pelican 111, 116, 192
Pennsylvania 186
Percival, A. 330
Perk, Prof K. 258
Persia 102
petrel 186
pig 87, 88, 89, 90, 91, 93, 186
Pitman 175
Pittsburgh Zoo 78
Pixie 41, 60, 63
Player, J. J. 163

poachers 13, 40, 62, 109-11, 122, 132, 134, 198, 265, 276, 322-3
Podoczak, Gerhard 14, 19, 30, 35
poison 62
Poland 124
Poles, W. 303
Pongola River 173
Pooley, A. 172, 173, 175
Poppleton, Frank (game warden) 58, 108, 310-11
Port Elizabeth 42, 277
Portuguese East Africa 319
poultry 88
Powell-Cotton, Major 264
Pretoria Zoo 319, 320
Principe 240, 246
pygmies 337
Pyrrhus, King 262
python molurus 78, 79, 90, 91
python reticulatus 78, 79

quagga 319
Queen Elizabeth National Park 29, 112, 134, 155, 164, 266, 278, 280, 284, 287, 291, 294, 310, 313

rabbits 87, 134, 141, 285
Radford, Col. 283
Ramerberg am Inn 179
Raswan 255
rat 87, 92, 263
   desert 255
   kangaroo 256, 258, 300
ratel 69
raven 229
Raven, Henry 93
redbreast 180
Red Sea 14
reed buck *see* antelope
reindeer 60
Resochin syrup 20
Reynolds, Vernon 20, 21
rhinoceros 13, 18, 29, 30, 59, 67, 71, 110, 136, 197, 248, 283, 300, 316f, 335, 336, 340, 347, 348, 349
   African 39
   armoured, Indian 64
   at Amboseli 39, 41, 60, 63, 65
   at Ngorongoro 56
   black, hook-lipped 9, 38f, 267, 316f
   dummies 340, 346, 349-52
   environment/behaviour 41-2, 44, 49, 267, 318f

excretion 51, 52, 318
food 50-1, 267
horns 39-40, 42, 64, 318
hunted 42
Indricothericum asiaticum 38
in zoos 39, 51-2, 57, 58-9, 60, 64-5, 66
mating 57-60, 318-19
numbers 43, 49-50, 64, 319-21
parasites of 55, 65-6
physique 38-40, 317-18
pregnancy and birth 52-3
preservation and settlement of 44, 65, 267, 320, 322, 323
relationship with elephants 53-4
relationship with other animals 54, 71-2
released on Rubondo 35, 44
sleep 51-2
triple-horned 40
violence 41, 54, 56-7, 58, 60-3
white, square-lipped 38f, 116, 248, 316f
woolly 38
Rhinos belong to everybody (Grzimek) 18, 275
Rhodesia 43, 76, 143, 167, 266, 279-80, 294, 304, 319, 320
  Northern *see* Zambia
rice 14
Rift Valley 128, 278
rinderpest 318
Ringelnatz, Joachim 38
Ringling Brothers' Circus 246-7
Ritchie, A. 54
Riviera 134
roan antelope *see* antelope
Robert 14, 23, 30-3, 36
rodents 80, 89, 174
roedeer *see* antelope
Romans 100, 262
Rome 177
Roosevelt, Theodore 74
Root, Alan 112, 113, 121-2, 130-1, 190, 197, 200, 209-11, 348-52
Root, Joan 113, 122, 209-11
Roth, H. 90
Royal Society 187, 188
Rualas (Central Arabia) 255
Ruanda 44, 210, 229
Rubondo 12-13, 17, 18, 19-35, 44
Rudolf, Crown Prince 191
Rudolf, Lake 161, 169, 172, 190, 191, 197, 200, 206, 254
Rugorogota 328
Ruhe, Hermann 88, 168
Ruhengeri 281-2

Ruindi—Rutchuru Plains 266
Rukwa Game Reserve 154
Rukwa, Lake 161
Rukwa Valley 128
Rusermi 147-8
Russia 125
Ruvuma River 288
Ruvu River 161

sable 103
Sahara 42, 127, 129, 144, 156, 183, 254, 256, 260
St Louis Zoo 220
St Lucia, Lake 172, 173
St Philip's Mission 166
St Vincent 85
Salebabu Island 92
Same 63, 64
San Diego Zoo 79
sandpiper 163
Sanyati River 276
Sarasota 247
Saudi Arabia 102
Sauer, Eleonore 104, 105
Sauer, Franz 104, 105
Saurians 96 *see also* alligators, crocodiles
Scauras, M. Aemilius 177
Schaller, George B. 75, 209-12, 217, 224, 226, 229-30
Schaurte, Dr A. 40
Schenkel, Rudolf 51, 61-2
Scheren, Dr Brigette 337-8
Scherpner, Dr 58
Schiess, A. 283-4
Schillings, J. G. 65, 76
Schillings, K. G. 76
Schistocerca gregaria 125, 128
Schistocerca parenensis 125
Schmidt-Nielsen, Dr 256
Schmitt, Dr 225
Schitzer, Dr Edouard (Emin Pasha) 320-1
Schütt, Gerda 52
scorpions 233
Scotland 67, 186
sea-coral 338
sea-cow 261
sea-lion 186
Seitz, Dr Alfred 297
Selous, Frederick C. 43, 56, 75-6, 333
Semliki Nile 155, 176
  Valley 89
Seremunda, Lucas (game warden) 36-7

365

Index

Serengeti Research Institute 9, 209, 274
Serengeti National Park 10, 12, 15, 16, 17, 18, 35, 64, 68, 71, 75, 85, 94, 104, 130, 132, 133, 134, 135, 141, 142, 144, 149, 151, 152, 153, 166, 204, 207, 274, 282, 324-6, 328, 329, 331-2, 335, 351
Serengeti Plain 50, 178, 268, 326
Serengeti shall not die (Grzimek) 207
Seronera 35, 141, 151, 274, 328
Sese Islands 167
Sévigné, Madame de 93
sheep 106, 183
Shehe (game warden) 56
Sheldrick, David 289-90
Shenton, J. (game warden) 94
Shiftas 198
Siberia 261
Siblin 247
Siks, Sylvia K. 267-8
Silesia 124
silver fox 103
Simba (Guggisberg) 55
Simba 73
Simon, Noel 265
Siren (sea cow) 261
Situtungas 18
sleeping sickness 181
Smit, Koos 54
snail 174
snakes 9, 77f, 121-2, 178, 186, 300
    adder, puff 184
    anaconda 77, 78, 79, 89, 92, 94
    as food 93, 94
    as royal symbol 97
    bites 29, 121-3, 151-2
    boa constrictor 77, 79, 85
    boa 77, 78, 86, 91, 94, 98
    capturing 85
    cobra 93, 184, 186
    cobra, king 78, 80
    farm 186
    food 89
    hunt 86-93, 96-7
    in myths 97
    in zoos 78, 79, 87-8, 90, 91-2, 94-5
    mamba 123, 151, 184
    mamba, African 78
    mamba, black 90
    mating, breeding, young 79, 94-6
    movement 85, 86
    physique 77, 80, 86, 90, 93, 94-6
    python 29, 77f, 110, 144, 184-5, 233
    python, amethyst 78

python, Asiatic reticulated 78, 86, 88, 90, 91, 92
python, East Asian reticulated 78, 79
python, Indian 78, 79, 90, 91
python, rock 78-9, 80, 86, 89, 90, 94, 95, 96-7
rattlesnake 93
skin 78, 93
taming 88, 91-2
viper 93
viper, Gaboon 91
Solander, Dr Daniel 187-8
Somalia 39-42, 197, 263
Somali bandits 197, 198; natives 281
Somaliland 39, 263
Sous Valley 125
Spain 18, 67, 183
Spaniards 237f
sparrow 263
spiders 233
Spirastachys africanus 50
Spitz, R. A. 337
spur-winged plover *see* plover
squirrel, ground 161
stable fly 65
stags 56, 141, 151
Stanley, H. M. 155
starling 65, 180
Starunia 38
Stephenson, J. (game warden) 327, 333
Stevenson-Hamilton, J. (game warden) 166
Stigand, C. H. 317
storks 178, 338
    marabou 163, 172, 178
Suakin 107
subungulates 261
Such agreeable friends (Grzimek) 11
Sudan 43, 190f, 232, 275, 293, 316-17, 321, 335
Suez 91, 106-7
Sumatra 168
suricate 192
Switzerland 18, 64

Taberer (game warden) 72
Tanga 15
Tanganyika, Lake 19, 20, 109, 171, 235
Tanzania (formerly German East Africa, then Tanganyika) 10-13, 14, 15, 18, 42, 44, 49, 55, 63, 97, 128, 131, 148, 154, 161, 208, 225, 265, 278, 282, 283, 288, 295, 299, 324-5, 327, 332, 333, 347-52
    Game Department 265
    National Parks 61, 201

Index

tapeworms 66
Taunus 313
Tchad, Lake 42, 232
Teleki von Szek, Count Samuel 190-2, 197
Telski, Count 70
termites 30, 61, 104, 291
Terreblanche, Sandy 90
Thomas 237f
Thomson's gazelles *see* antelope
thorn-bush 50
Thulo 153
ticks 55, 65
Tiesenhausen, Baron von 179
tiger 68, 70, 71, 74, 134, 135
  white 70
'*Times of Swaziland*', the 166
Tiran 42-3
toad 174
Tokyo Zoo 154
Tooly, Mrs Noel 327
topi *see* antelope
Torit 201, 202, 204
Toro Game Reserve 89
tortoise, aquatic 54-5
  giant 187
Toto 212, 233-53
tourists 18, 56, 63, 70, 71, 76, 90, 101-2, 109, 110, 116, 121, 123, 136, 141, 143, 152, 153, 156, 198, 275, 322, 327
Transvaal 42, 147, 316, 319
  Snake Park 94, 95
Trochilus 163
Trubka 136
Tsavo National Park 51, 56f, 136, 154, 168, 183, 184, 248, 266, 267-8, 273, 289-90, 292
tsetse fly 70, 181-2, 320
Tschombe, M. 205
Tuaregs 127
Tunduru 288
Turkana 190-2, 281, 294
Turkey 183
Turner, Myles (game warden) 35, 44, 142, 144, 166
turtles 172
Twenty animals, one man (Grzimek) 218

Ubangi 42, 321
Uganda 29, 53, 89, 109-10, 111, 112, 134, 144, 148, 155, 161, 176, 184, 190, 198, 199, 200, 264, 265, 266, 276, 280, 286-7, 288, 294, 298, 300, 317, 320, 321, 328
  Game Department 265
  National Parks 121, 295, 297

Railway 75
Uganda Kob *see* antelope
Ukerere (Island) 70
Umfolozi Black River 319
  Game Reserve 318-20
  White River 319
Ungulates 261
U.N.O. 191-2, 208, 316
United States of America 10, 18, 64, 93, 109, 181, 262, 320
Usutu River 166-7
Utu 227

Van der Merwe, Dr N. 300
varan, Nile 170, 172-3
Venice (Florida) 247
Verschuren, Dr 293
Victoria (Australia) 184
Victoria Falls 155
Victoria, Lake 12f, 29, 44, 70, 161, 162, 167, 190, 204, 208
Victoria Nile *see* Nile, Victoria
Vienna Zoo 70, 334
Voeltzkow 171
Volkswagen 15, 35, 131, 279, 336, 352
vulture 74, 89, 94, 105, 130, 165, 172, 293, 310

Wagner, Dr H. O. 183
Wakamba 62
Wakambaland 43
Wales 67
Wally 241
Walther, Dr Fritz 17, 30, 31-5
Wankie Game Reserve 284
  National Park 266, 284, 300, 320
Warmbrunn, Elizabeth 179-80
warthog 144, 181, 182, 284, 325, 326
Washington Zoo 79, 95
waterbuck *see* antelope
Watson, Mr 49
Watusi 210
Webb, George 334
Wehlitz, Capt. K. W. 14, 15
Wells, Mr 74
West Africa 16, 20, 42, 167, 209, 232, 264, 274
Wiesbaden 205
Wikingen, Dr 136
wild dog 69, 133, 144, 268, 324f
  African 331
  extermination of 324
  grouping 325-9
  hunting 324-9
  'hyena-dogs' 324

367

Index

wild dog [contd.]
  in zoos 329, 334–5
  mating, birth, etc. 330–2
  speed 325
Wimpy 350
witchcraft 187
witch doctor 168, 314
Wogakuria, G. W. 351
Wolhuter (game warden) 73
wolves 11
World War I 99, 106, 133, 164
worms 65–6
  filarial 65
  round 164
  tape 66
Wright, Allan 309
Wright, Bruce S. 74, 326

Young, Lake 40
Yugoslavia 183

Zambia (formerly Northern Rhodesia) 40, 41–2, 43, 90, 154, 161, 165, 182, 276, 289, 295, 303, 311, 313, 326, 327, 328
  legend 51
Zambezi River 42, 263, 266, 294, 319
Valley 276, 294
Zanzibar 265
zebra 18, 35, 44, 54, 69, 74, 75–6, 131, 132, 133, 136, 141, 142, 166, 178, 181, 185, 197, 268, 300, 334, 336, 340
  dummies 72
  relationship with other animals 72
zebu 185
'Zoological Society of 1858' (Frankfurt) 13
Zoological Society of Frankfurt 121
Zoological Society of New York 78
zoos 16, 22, 23, 36, 39, 57, 58, 64, 66, 237, 320, 330–1; *and see* American 262, Barcelona, Basle, Berggreen, Berlin, Berne, Boras, Breslau, Bristol, Bronx, Chicago, Columbus, Copenhagen, Cros de Cagne, Dresden, European, Feruviks Parken, Frankfurt, Hanover, Khartoum, London, Manchester, Munich, Nairobi, New York, Nuremberg, Opel, Paris, Pittsburg, Pretoria, Rio de Janeiro, Rome, St Louis, San Diego, Tokyo, Vienna, Washington
Zululand 64, 163, 172, 173, 174–5, 176, 197, 248, 317, 319
Zurich 67

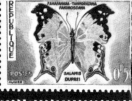